Roland Conrady · Martin Buck
Editors

Trends and Issues in Global Tourism 2008

Springer

Professor Dr. Roland Conrady
University of Applied Sciences
Erenburgerstr. 19
67549 Worms
Germany
conrady@fh-worms.de

Dr. Martin Buck
ITB
Messedamm 22
14055 Berlin
Germany
buck@messe-berlin.de

ISBN 978-3-642-09653-2 e-ISBN 978-3-540-77798-4

DOI 10.1007/978-3-540-77798-4

Cover-design: WMX Design GmbH, Heidelberg

Printed on acid-free paper

9 8 7 6 5 4 3 2 1

springer.com

Preface

As other industries, the global travel and tourism industry has been facing immense challenges and highly visible upheaval since the beginning of the new millennium. The International Tourism Exchange ITB Berlin, the world's leading travel trade show, aims at pinpointing the most important challenges, identifying the trends and offering a platform to solve pressing problems. The ITB Convention Market Trends & Innovations has developed into a center of excellence and a driving force for the global travel and tourism industry, generating a much needed information platform.

This compilation unites the highlights of the convention in articles prepared by renowned professionals and scientists from the industry. Readers may benefit from this comprehensive vision of the developments, which are shaping the structure of the global tourism industry today and in the future. This book is indispensable for tourism and travel professionals as well as for academics and students analyzing current global tourism and travel trends.

With the latest empirical data on the world population's travel behavior and the topic of the year 2007, climate change, the first chapter provides the basis for this book. With over one million interviews per year in 57 countries, the World Travel Monitor presents the largest travel survey worldwide. With their article, Rolf Freitag and Dennis Pyka provide insights into global travel patterns and trends. There are hardly any other issues that concerned the global travel and tourism industry in 2007 as much as global climate change. Roland Conrady and Stephan Bakan outline the latest findings on climate change and point out the implications for the tourism industry. Clearly defined measures for the strategic options adaptation and prevention are presented.

The second chapter deals with two aviation topics, which are increasingly thrust into the spotlight. Raimund Hosch describes the business aviation market segment. He identifies the potential of this segment and draws the conclusion that due to novel, more economical aircraft and innovative business models, the business aviation segment is up to an accelerated growth. Adrian von Dörnberg presents the status quo on airport privatizations and takeovers worldwide. Driving forces of the global airport business are analyzed and it is derived to what extent airport privatizations and takeovers contribute to the achievement of the different airport stakeholder objectives.

The third chapter is dedicated to hospitality management. Interviewed by Maria Pütz-Willems, designer Matteo Thun pinpoints trends in hotel architecture. In addition he explains the business model of his new hotel chain ‚Zero Star', which will cause a furor in the hotel industry. Werner Pauen analyzes a financially delicate subject of the hotel industry. He demonstrates that the existing analysis methods on credit ratings in the hotel industry are unsuitable and presents a new, advanced model. Susanne Fittkau and Axel Jockwer address the phenomenon that in times of Web 2.0 booking decisions increasingly rely on the hotel ratings of other Internet users. A case study on community website HolidayCheck illustrates how online hotel ratings work and what effects they have.

The fourth chapter deals with the effective and efficient management of tourist destinations. Jürgen Ringbeck, Stephan Gross, Thea Chiesa and Jennifer Blanke introduce results and recommendations for action of the first Travel & Tourism Competitiveness Report by the World Economic Forum. The report, which has analyzed the competitiveness in the tourism and travel sector of 124 nations, supports decision makers in politics in order to strengthen an industry that is particularly suited to boost economic development and the level of prosperity in a country. Harald Pechlaner, Petra Hedorfer and Norbert Tödter address the most important tourism destination worldwide: Europe. Due to an increased competitiveness of destinations in Asia, European destinations feel compelled to further develop and improve their product and brand management. The authors pay particular attention to cooperation strategies and arrangements and comment on Germany as an example. One of the most important emerging markets is India. This country offers great potential for the development of religious and medical tourism. John Koldowski and Oliver Martin show how India is able to satisfy two travel incentives that are gaining in importance worldwide.

The fifth chapter covers contemporary issues in marketing and sales management. Hans Rück and Marcus Mende attend to the mega trend of individualization of societies. Classical approaches of market segmentation therefore reach their boundaries. The authors introduce an innovative segmentation method, which makes it possible to address individual target groups. Derived from empirical studies, Dan Greaves describes a vision of future traveler tribes in the year 2020. Particular attention is paid to the utilization of new information technologies and the resulting implications for airlines. Using the example of a travel agency cooperation, Walter Freyer and Manuel Molina demonstrate innovative destination marketing practices. Modern travel agency cooperations allow for destinations to market their products and services in a multi-channel system.

Chapter six provides a survey of online travel in Europe and business travel in Germany. With empirical data, provided by PhoCusWright Inc., Michaela Papenhoff, Klaus Fischer and Roland Conrady analyze the structure of an already gigantic online travel market in Europe. The authors display the challenges for online travel agencies, airlines and tour operators. Kirsi Hyvärinen and Michael Kirnberger provide a deeper insight into a due to its size often misjudged market segment. With close to 50 billion Euros the German business travel market is of almost the same magnitude like the German private travel market.

Closing, the seventh chapter deliberately adds a new angle. Tony Wheeler recaptures his travel experience of 35 years as well as his experience as founder of Lonely Planet. His evaluation of new market segments and new destinations is always inspiring and sometimes even provoking. In any case they are of considerable value for the development of innovative travel products. More intense travel experiences than when traveling to space is close to impossible. Dana Ranga interviewed the Head of the Astronaut Division at the European Space Agency (ESA), Gerhard Thiele, and found out what men are feeling when on the way to new horizons. This knowledge is a crucial basis for the understanding of a very central travel motive: the discovery of the unexplored and the collection of extraordinary experiences.

This work could not have been achieved without the remarkable dedication on behalf of the authors, who, for the most part, have taken on executive positions in the tourism economy. Special thanks go to Daniela Rist, Assistant at the Faculty of Tourism and Travel, University of Applied Sciences Worms. With her eagerness, great skill and best care she contributed significantly to a successful publication. Also, without Alexandra Salamon and Alexandra Saless, both former students from tourism and travel faculties, this publication would have hardly been possible.

Frankfurt/Berlin, January 2008

Roland Conrady
University of Applied Sciences Worms

Martin Buck
Messe Berlin

Contents

Key Figures and Issues in the Global Tourism Industry

Global Tourism in 2007 and Beyond –World Travel Monitor`s Basic Figures

Rolf Freitag, Dennis Pyka

Climate Change and Its Impact on the Tourism Industry

Roland Conrady, Stephan Bakan

Aviation Management

The Role of Business Aviation in the European Civil Aviation Market

Raimund Hosch

Airport Privatization and Takeover – Creating Value for All Stakeholders?

Adrian von Dörnberg

Hospitality Management

Hotel Architecture: Less Is More - Or: Natural Sustainable

Maria Pütz-Willems

Financial Solvency and Credit Rating for Hotel Businesses

Werner Pauen

Quality Rating in Hotel Community Sites

Susanne Fittkau, Axel Jockwer

Destination Management

Improving Travel & Tourism Competitiveness

Jürgen Ringbeck, Stephan Gross, Thea Chiesa, Jennifer Blanke

European Destination Management: Challenges for Product and Brand Management

Harald Pechlaner, Petra Hedorfer, Norbert Tödter

Emerging Market Segments: Religious and Medical Tourism in India

John Koldowski, Oliver Martin

Marketing and Sales Management

Innovations in Market Segmentation and Customer Data Analysis

Hans Rück, Marcus Mende

Customer Segmentation Traveller Types and Their Needs in 2020

Dan Greaves

Innovations in Destination Distribution Management

Walter Freyer, Manuel Molina

Travel Technology and Business Travel Management

European Online Travel Overview – An Abstract of PhoCusWright Online Travel Overview Third Edition

Michaela Papenhoff, Klaus Fischer, Roland Conrady

Trends in the Business Travel Market – Focus on Germany

Kirsi Hyvärinen, Michael Kirnberger

The Future of Tourism and Travel

Future Destinations

Tony Wheeler

Space Travel

Gerno Reiter

Authors

Bakan, Stephan, Dr.
Max Planck Institute for Meteorology
Bundesstr. 53, 20146 Hamburg
Germany

Blanke, Jennifer
Senior Economist
World Economic Forum
91-93, route de la Capite, 1223 Cologny/Geneva
Switzerland

Buck, Martin, Dr.
Director
Messe Berlin GmbH
Competence Centre Travel & Logistics
Messedamm 22, 14055 Berlin
Germany

Chiesa, Thea
Head of Aviation, Travel & Tourism
World Economic Forum
91-93, route de la Capite, 1223 Cologny/Geneva
Switzerland

Conrady, Roland, Prof. Dr.
University of Applied Sciences Worms
Department of Tourism and Travel
Erenburgerstr. 19, 67549 Worms
Germany

Fischer, Klaus, Prof. Dr.
University of Applied Sciences Worms
Department of Tourism and Travel
Erenburgerstr. 19, 67549 Worms
Germany

Fittkau, Susanne
CEO
Fittkau & Maaß Consulting GmbH
Hohe Bleichen 28, 20354 Hamburg
Germany

Freitag, Rolf
CEO
IPK International World Tourism Marketing Consultants GmbH
Gottfried-Keller-Straße 20, 81245 Munich
Germany

Freyer, Walter, Prof. Dr.
Dresden University of Technology
Faculty of Transportation and Traffic Sciences "Friedrich List"
Andreas-Schubert-Straße 23,
Zi. 520, 01062 Dresden
Germany

Greaves, Dan
Global Head of Travel Agency Communication
Amadeus
Salvador de Madariaga 1, 28027 Madrid
Spain

Gross, Stephan, Dr.
Senior Associate
Booz Allen Hamilton
Lenbachplatz 3, 80333 Munich
Germany

Hedorfer, Petra
Chairwoman of the board
Deutsche Zentrale für Tourismus (DZT)
German Tourist Board
Beethovenstr. 69, 60325 Frankfurt
Germany

Hosch, Raimund
CEO
Messe Berlin GmbH
Messedamm 22, 14055 Berlin
Germany

Hyvärinen, Kirsi
Senior Specialist
BearingPoint GmbH
Gladbecker Str. 5, 40472 Duesseldorf
Germany

Jockwer, Axel, Dr.
Head of Marketing Department
HolidayCheck AG
Romanshornerstrasse 115, 8280 Kreuzlingen
Switzerland

Kirnberger, Michael
President
Verband Deutsches Reisemanagement e.V.
The Business Travel Association of Germany
Darmstädter Landstraße 110, 60598 Frankfurt
Germany

Koldowski, John
Director - Strategic Intelligence Centre
Pacific Asia Travel Association (PATA)
Unit B1, 28th Floor, Siam Tower, 989 Rama
Road, Pathumwan, 10330 Bangkog
Thailand

Martin, Oliver
Associate Director - Strategic Intelligence Centre
Pacific Asia Travel Association (PATA)
Unit B1, 28th Floor, Siam Tower, 989 Rama
Road, Pathumwan, 10330 Bangkog
Thailand

Mende, Marcus, Dr.
CEO
Schober Information Group Deutschland
Max-Eyth-Straße 6-10, 71254 Ditzingen
Germany

Molina, Manuel
CEO
Touristik Service System GmbH
Budapester Str. 34b, 01069 Dresden
Germany

Papenhoff, Michaela
Managing Director, Europe
PhoCusWright
1 Route 37 East, Suite 200, Sherman, CT 06784-1430
United States

Pauen, Werner
Bad Tölz
Germany

Pechlaner, Harald, Prof. Dr.
Katholische Universität Eichstätt-Ingolstadt
Institute for Management and Tourism
Pater-Philipp-Jeningen Platz 2 / 1. Stock, 85072 Eichstätt
Germany

Pütz-Willems, Maria
Editor-in-Chief
hospitalityInside.com
Paul-Lincke-Str. 20, 86199 Augsburg
Germany

Pyka, Dennis
IPK International World Tourism Marketing Consultants GmbH
Gottfried-Keller-Straße 20, 81245 Munich
Germany

Ranga, Dana
Director
Good Idea Films
Motzstr. 88, 10779 Berlin
Germany

Ringbeck, Jürgen, Dr.
Senior Vice President
Booz Allen Hamilton
Zollhof 8, 40221 Duesseldorf
Germany

Rück, Hans, Prof. Dr.
University of Applied Sciences Worms
Department of Tourism and Travel
Erenburgerstr. 19, 67549 Worms
Germany

Thiele, Gerhard
Head of the ESA Astronaut Division ESA/EAC
Linder Höhe, 51147 Cologne
Germany

Thun, Matteo
Matteo Thun & Partners S.r.l.
Via Appiani 9, 20121 Milan
Italy

Tödter, Norbert
Head of Strategic Planning and Market Research
Deutsche Zentrale für Tourismus (DZT)
German Tourist Board
Beethovenstr. 69, 60325 Frankfurt
Germany

von Dörnberg, Adrian, Prof. Dr.
University of Applied Sciences Worms
Department of Tourism and Travel
Erenburgerstr. 19, 67549 Worms
Germany

Wheeler, Tony
Founder of Lonely Planet
Lonely Planet
90, Maribyrnong Street, 3011 Footscray, Victoria
Australia

Key Figures and Issues in the Global Tourism Industry

Global Tourism in 2007 and Beyond – World Travel Monitor's Basic Figures

Rolf Freitag, Dennis Pyka

1 Introduction

This abstract World Travel Trends Report 2006-07 is based primarily on the 2006 results of IPK International's World Travel Monitor – the continuous tourism monitoring system, which was first set up in 1989 and now undertakes more than 1 million interviews a year in 57 of the world's major outbound travel markets – 36 across Europe and 21 from the rest of the world. The interviews are designed to be comparable from one year and from one market to another, and to yield information on market volumes and sales turnover, destinations, travel behaviour, motivation and satisfaction, travellers and target groups, and recent tourism trends.

IPK International's ITB Berlin Message, presented at the ITB Convention during ITB 2007, focused on World and European trends, as well as looking in more detail at the German travel market, domestic and outbound. This report provides more detailed information than it was presented in Berlin, drawing on information from other sources, such as the World Tourism Organization (UNWTO), as well as markets included in the World Travel Monitor but not presented in Berlin.

2 Overview of World Tourism in 2006

2.1 Global Trends

The world economy is riding the crest of a wave, and the world travel and tourism industry is travelling with it. Global GDP increased at its fastest rate in a quarter of a century in 2004-06, and so far – in spite of occasional worries – there has been little sign of a significant slowdown.

According to preliminary results published by UNWTO, international tourist arrivals increased by 4.5% to 850 million in 2006. IPK International's World

Travel Monitor, which measures outbound travel, points to an increase of 5% to 650 million overnight trips worldwide. (The difference in respective volumes can be explained by the fact that one outbound trip can involve travel to several destinations, resulting in a higher count for cumulative arrivals worldwide.)

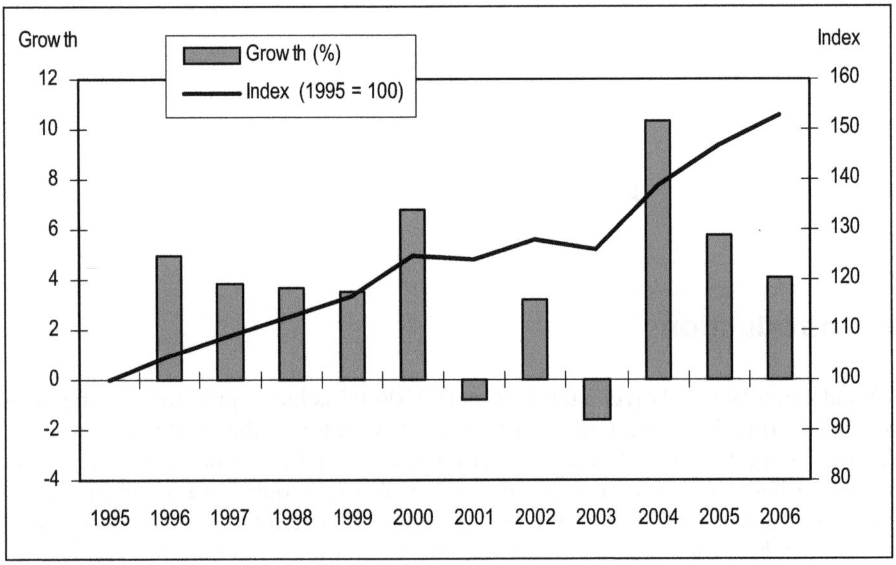

Fig. 1: World tourism performance, 1995-2006; Sources: World Tourism Organization (UNWTO); World Travel Monitor, IPK International

2006 was therefore another good year for international tourism. However, in a sense it also represented a return to more normal patterns of growth. After the troubles of 2001-03, 2004 was dubbed the year of 'super recovery' by IPK. It brought international travel back close to the level implied by its medium-term growth trend, as figure 1 shows. And in 2006 growth settled back close to that medium-term trend – at 4-5% a year.

Tourism continued to face difficulties last year, including occasional terrorist attacks, and threats of attacks, on tourists in Turkey, Egypt, the UK and other countries, the panic over bird flu, increased fuel prices, the usual crop of storms, earthquakes and other natural disasters, and the traditional complaints about poor weather – a damp summer in Central and Northern Europe, for instance, and a lack of snow in many winter resorts.

However, the year did not turn out to be the disaster forecast by much of the media during ITB Berlin 2006. It was not marked by any real catastrophe as in 2001-04 – 9/11 in 2001, the SARS epidemic in 2003, and the tsunami of December 2004 – and there was nothing like the economic downturn brought on by the end of the dot.com boom in 2000. Most importantly, the fears of an avian flu pandemic were not (yet) realised.

On the positive side, there has been a dramatic change in the pattern of global economic growth. There is a much greater focus on international trade – to the benefit of international tourism. And there has been a huge shift in terms of trade – and therefore in purchasing power – in favour of commodities producers. These changes are reflected in the regional trends in tourism growth.

2.2 Inbound Performance

There were greater regional, and sub regional, disparities in growth in 2006 than in 2005. The increase in arrivals in the Americas, Europe and the Middle East slipped back to 2%, 4% and 4% respectively, while arrivals in Asia Pacific increased to 8% – reflecting the strong recovery of most destinations directly affected by the December 2004 Indian Ocean seaquake and tsunami. Meanwhile, the growth in arrivals in Africa remained encouragingly high, at 8%.

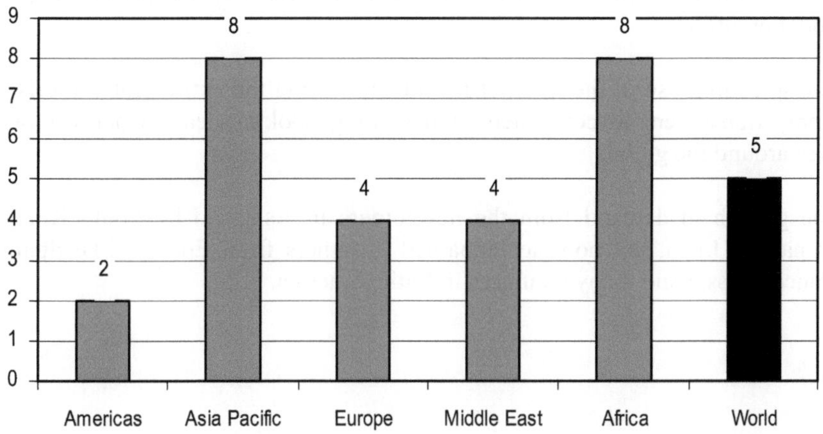

Fig. 2: World tourism performance by region, 2006 (% annual change in international arrivals); Sources: UNWTO; World Travel Monitor, IPK International

In Asia Pacific, Northeast Asia, Southeast Asia and South Asia all put in excellent performances, driven by the emergence of China and India as leading destinations and sources of travelers. However, there was no growth in arrivals in Oceania, marking the end of a period of rapid growth. This was due in part to the strength of the Australian and New Zealand dollars against the US dollar, but also to the fact that these relatively remote destinations – Australia, New Zealand and the Pacific Islands – suffered from fuel surcharges on airfares.

North Africa and (despite regional conflicts) the Middle East both did reasonably well in 2006, but SubSaharan Africa again led the region.

Western and Southern Europe achieved 4% increases in arrivals – sound performances for mature markets – and Northern Europe an exceptional 7%, with strong performances from Scandinavia, the UK and Ireland. Tourism in Eastern Europe took a rest.

Most of the growth in the Americas was contributed by Central and South America; the growth in arrivals in North America and the Caribbean was much weaker. US travellers, in particular, remain wary of international travel and have likely been discouraged by the weak dollar.

IPK International has identified four main drivers of global travel and tourism growth:

- The increased wealth, which stimulates demand involved in the global economic boom;

- Falling prices due to ongoing gains in tourism productivity, reflected in rapidly spreading low-fare flight offers and, more recently, low-priced accommodation offers;

- The spreading use of the Internet for travel, making the entire global tourism supply transparent to consumers, and making bookings easy wherever one lives around the globe;

- The growth in demand from the newcomers in Asia, led by South Korea, China and India, but now accompanied by others from Europe, like Spain, France, Russia and many countries in Latin America.

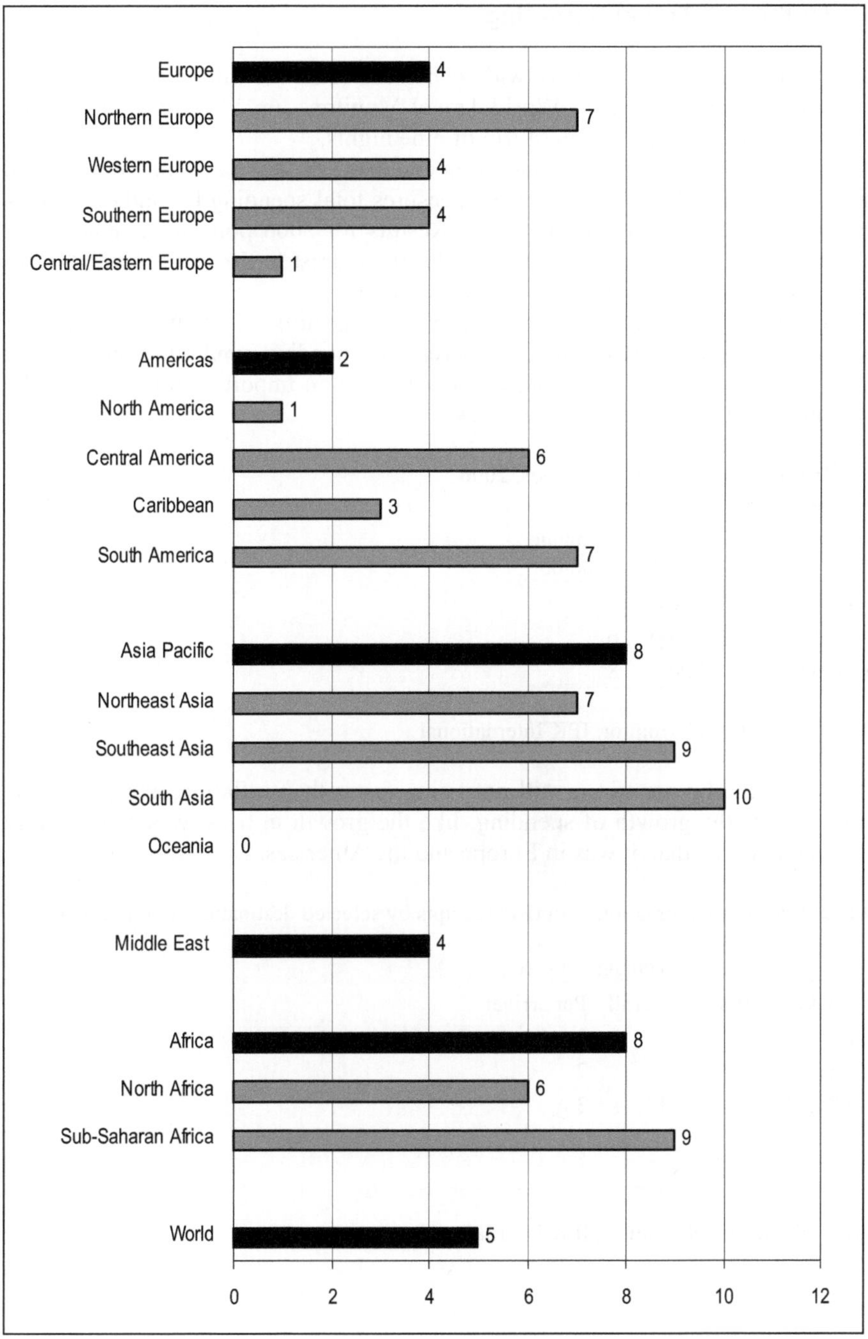

Fig. 3: Growth in international tourist arrivals by region and sub-region, 2005 (% change over previous year); Sources: UNWTO; World Travel Monitor, IPK International

2.3 Outbound Travel Spending

The 650 million trips taken worldwide generated some 6.0 billion overnights, according to IPK International's World Travel Monitor – up 5% over 2005's level. This suggests an average length of trip of nine nights.

Unlike UNWTO, which uses data from countries' central banks on international tourism receipts and expenditure, IPK measures total spending by outbound travellers – before as well as during a trip. Results for 2006 point to an 8% rise in spending to €700 billion, or €1,100 per trip. In contrast to the general trend in recent years, spending outpaced both trips and nights.

Note that in figure 4 the spending is expressed in Euro – a relatively strong currency in 2006. In US dollar terms, the increase in spending would have been rather higher. The weakness of the dollar continues to be an important influence on the selection of destinations around the world.

Table 1: World outbound travel trends, 2006

	Total	%change 2006/05
Trips (mn)	650	5
Nights (mn)	6,000	6
Spending (€ bn)	700	8

Source: World Travel Monitor, IPK International

However, spending per trip is still not rising faster than inflation, except in Asia Pacific, where the growth of spending, like the growth in trips, was significantly faster in real terms than it was in Europe and the Americas.

Table 2: Growth in international tourism receipts by selected destination region, 2006/05

	% change in receipts	
Destination region	Overall	Per arrival
Europe	5	2
Americas	5	2
Asia Pacific	11	3
World	8	2

Source: World Travel Monitor, IPK International

3 European Travel and Tourism Trends

3.1 Overall Travel Demand

In 2006 European adults of 15 years and over made 389 million trips abroad of a minimum one night's stay, according to IPK International's European Travel Monitor. These generated 3.7 billion nights and €354 billion in spending.

Trip volume increased overall by 3% in 2006, the European Travel Monitor shows. There was also a welcome 2% increase in average length of trip, after declines of 4% in both 2004 and 2005. This resulted in a 5% rise in overnight volume.

Table 3: European outbound travel, 2006

	2006	%change on 2005
Trips[a] (mn)	389	3
1-3 nights long	103	6
4+ nights long	286	2
Overnights (mn)	3,700	5
Average length of trip (nights)	10	2
Spending (€ bn)	354	6
Spending per trip (€)	898	1
Spending per night (€)	95	1

[a] Trips made by adults aged 15 years and over
Source: European Travel Monitor, IPK Interna- tional

The trend towards shorter but more frequent trips is nevertheless well entrenched. There was a 6% increase in trips of one to three nights, compared with a rise of only 2% in trips of four nights and longer.

Meanwhile, although the estimated increase in total expenditure by Europeans on travel abroad was 6% – compared with 5% in 2005 and 4% in 2004 – this translated into increases of just 1% in spending per trip and 1% in spending per night – slightly less than average inflation in Europe and considerably less than average inflation in most other destinations.

More detailed analyses would no doubt reveal some interesting trends. Given the increased take-up of low-fare flights on short-haul routes and the decline in accommodation prices in some destinations, there is scope within the overall figures, shown in table 3, for an increase in real spending in short-haul destinations. But the increases in long-haul airfares (related mainly to higher fuel prices) may also have distorted the overall figures.

Holidays account for more than two thirds of total European outbound trips (69%) while business travel, which had lost share over recent years, but which recovered in 2005, recorded a second consecutive year of stronger than average growth, taking its share to about 15% – up from just 11% in 2004. Visits to friends and relatives (VFR) and other leisure trips account for some 17% of total trips, a two-point loss over 2004's level.

3.2 Major Source Markets and Destinations

There has been little significant change in the ranking of the leading European source markets. The most important outbound markets in Europe are, in order of size, Germany, the UK, France, the Netherlands and Italy. Together, the top five accounted for 58% of total trip volume in 2006.

Table 4: Leading European outbound travel markets, 2006

Market	No. of trips (mn)	% change 2006/05
Germany	74.5	-2
UK	62.5	2
France	29.7	4
Netherlands	21.5	1
Italy	21.1	8
Russia	15.9	10

Source: European Travel Monitor, IPK International

Of the top five, Germany was the only one to suffer a decline in outbound travel demand (-2%). The UK continued its unbroken record of annual growth, driven in large part by low-cost/low-fare travel, and both France (+4%) and especially Italy (+8%) showed above average growth. However, now in sixth position in the overall ranking, Russia outpaced all the leaders with a 10% increase – following a 14% rise in 2005 and 17% in 2004.

Two other countries in Central/Eastern Europe – Poland and the Czech Republic – ranked among the fastest growing European markets in 2006, while Spain (+8%) and Switzerland (+7%) are also growing faster than most. In the case of Spain and Italy, this level of growth is (finally) bringing them closer to the volumes warranted by their wealth and large populations.

Ireland's presence among the fastest growing markets is scarcely surprising, given the country's economic growth in recent years and the widespread availability of low-cost/low-fare flights. And Switzerland's rebound, after several years of stagnation and even decline, is attributable to a rise in consumer confidence and disposable incomes. Nevertheless, this level of annual increase may be short-lived because of the weakness of the Swiss franc, which has made foreign travel more expensive.

Table 5: Leading European growth markets for outbound travel, 2006

Market	% growth 2006/05
Poland	17
Russia	13
Ireland	10
Spain	8
Italy	8
Czech Republic	7
Switzerland	7

Source: European Travel Monitor, IPK International

The last four months of 2006, September through December, showed the strongest growth in outbound trips by Europeans last year, partly due to a base effect (this period showed the slackest growth in 2005), but probably also reflecting the increase in economic prosperity and consumer confidence in the euro area in the second half of the year.

Table 6: European outbound travel growth, 2006, by season

Season	% annual change in no. of trips
Jan-Apr	3
May-Aug	2
Sep-Dec	5

Source: European Travel Monitor, IPK International

As far as holiday travel is concerned, there were mixed performances from one segment to another in 2006. Sun & beach holidays increased by 4% – up from 3% in 2005 – maintaining its dominance of the European holiday travel market by a very wide margin. City breaks, which for years have been the main beneficiaries of the boom in low-cost/low-fare fights, now seem to be growing more gently – they were up 5% to around 42 million in 2006. Countryside trips did well (+9%), reflecting the growth in demand for holidays in rural surroundings, but events did best of all – attracting a growth of 14% – no doubt with the help of the FIFA World Cup in Germany and other mega events such as the year-long celebrations of Rembrandt's 400th anniversary in the Netherlands and the commemoration of the 250th anniversary of Mozart's birth in several European countries, but primarily in Austria and Germany.

Table 7: European outbound holiday travel growth by type of trips, 2006

Type of trip	% annual change in no. of trips
Events	14
Countryside	9
City breaks	5
Sun & beach	4

Source: European Travel Monitor, IPK International

Air travel continued to increase its share of the European outbound travel market in 2006. Trips by air increased by 7%, as against a 1% decline in trips by bus/coach and car. Travel by ship/boat rose by 8% in terms of trip volume, reflecting the fashion for cruises, with the Baltic Sea benefiting strongly from the trend. Trips by train on the other hand were flat, in spite of the improved high-speed rail services, including those through the Channel Tunnel.

The growth in low-fare air travel continues, but the rate of growth is moderating – as described in detail later in this report.

The top eight destinations for European travellers are all within Europe, although the USA – the favourite single long-haul destination – follows fairly closely behind the leaders. The ranking of the top eight remained unchanged in 2006, but France continued to lose ground. Turkey took a hit, primarily as a result of terrorist attacks, but also due to the incidence of bird flu in the country and, perhaps even more significantly, fear of reprisals against Western tourists following the publication of the Danish caricatures early in the year. More generally, sun & beach destinations in the eastern Mediterranean suffered from the regional conflicts, to the benefit of tried and tested destinations in the western Mediterranean.

Table 8: Top destinations of European outbound travelers, 2006

Rank	Destination	Trips (mn)	% change on 2005
1	Spain	50.6	4
2	France	39.9	-1
3	Germany	33.6	8
4	Italy	30.2	6
5	UK	21.3	8
6	Austria	20.7	1
7	Turkey	14.1	-10
8	Greece	12.6	5

Source: European Travel Monitor, IPK International

Although IPK has not released detailed trends, long-haul destinations performed well out of the European market in 2006. Demand for travel to the USA was down in 2005, but some markets showed good growth. These included Spain and Switzerland. Latin America, meanwhile, gained market share out of Europe. But the best-performing region was Asia Pacific. This was the result of the market's recovery following a sluggish year in 2005, due to the impact of the December 2004 tsunami.

2005's poor performance was not entirely due to a drop in demand for the tsunami-afflicted destinations, however. Following the tsunami, airlines reduced flight and seat capacity on different routes, anticipating a big drop, so that when the recovery started, there was not enough capacity to meet demand. This was especially true for the Maldives and for the island of Phuket in Thailand. As a result, there was a lot of pent-up demand at the beginning of the winter season 2005/06.

3.3 Booking Patterns

As far as the organization of travel was concerned, traditional travel agencies continued to lose ground in 2006. Europeans made 4% more advanced bookings in 2006 than in 2005, but 3% fewer through travel agencies. (Advanced bookings through the internet were up 21%, those 'direct to transport' were up 4%, and those 'direct to accommodation' were up 11%.) Trips involving no advance booking were flat.

All this is part of the continued growth in use of the Internet for travel – a trend which is maturing, and in which the headline growth rates are therefore moderating. Use of the Internet for online booking as opposed to simply 'looking' – gathering information prior to booking a trip – increased by 16% in 2006, following a 35% rise in 2005 and a 39% rise in 2004. This means that nearly one in three outbound trips now involves online booking for at least part of the trip. The equivalent share was just 19% in 2003.

Table 9: European online travel, 2006

	% growth 2006/05	% market share 2006
Bookers	16	32
Lookersa	-5	13
All online travel	9	45

[a]Use of the internet to research travel options, but not for booking
Source: European Travel Monitor, IPK International

Clearly, the incidence of Internet bookings among the smaller, less mature European markets is much lower, but it is growing fast – much faster, in fact, than online travel 'looking'. Although the travel trade still dominates, as far as travel distribution is concerned, around 44% of trips involved new distribution channels, e.g. online bookings through tour operators, travel agents or direct with suppliers.

The total number of trips involving online booking in 2006 was about 124 million. Of these, more than two thirds involved some kind of booking for accommodation and 60% for flights. Other products are much less likely to be booked online, either because the facility for such bookings is less widely available, or because Europeans are more nervous of booking these products through the Internet. Travel insurance is one obvious example.

The low share of rail travel in total online bookings is perhaps surprising, especially given the sophisticated technology in some national railway booking systems, such as the SNCF in France. But this is probably attributable to the fact that markets like France rarely opt for train travel when going abroad. It is much more popular for domestic trips.

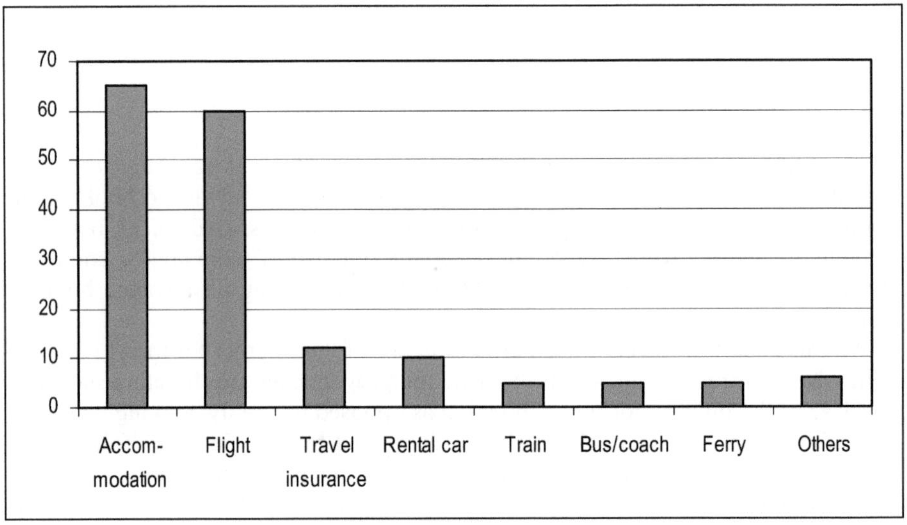

Fig. 4: Online bookings by European outbound travellers, 2006 (% breakdown of items booked by European online travellers[a]); Source: European Travel Monitor, IPK International ([a] Multiple responses possible)

4 German Travel & Tourism

4.1 Inbound Trends

German inbound travel recorded three very strong years of growth in 2004-06, after the vicissitudes of 2001-03. In 2006, according to official data, Germany attracted 23.6 million foreign visitors (measured at all forms of commercial accommodation establishments), up 9.5% after increases of 9.4% in 2004 and 6.8% in 2005. By comparison, Europe as a whole managed increases of only about 4% in those three years. The 9.5% increase in 2006 was, of course, heavily indebted to the FIFA World Cup hosted by Germany during the months of June and July. The second quarter of 2006 saw a 14.5% increase in arrivals, compared with the same quarter of 2005.

Germany's international tourism receipts rose by 8.9% in 2004, 5.5% in 2005 and 11.6% in the first eleven months of 2006, in euro terms – including a 19% increase in the second quarter.

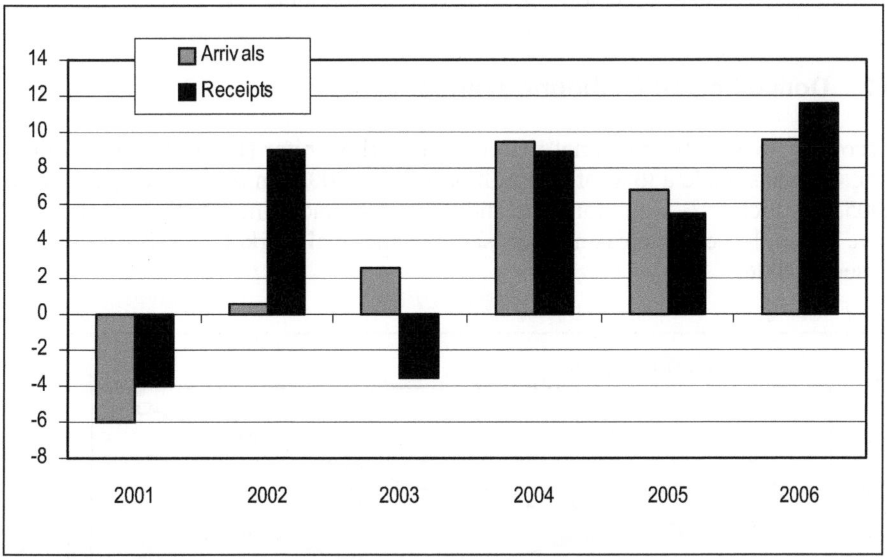

Fig. 5: Trends in German inbound travel, 2000-06 (% annual change[a]); Source: Deutsche Bundesbank; ([a]Arrivals measured at all forms of commercial establishment; receipts (for all visitors) in euro terms)

Foreign nights in commercial establishments in Germany – hotels and other forms of paid accommodation – were up 15% to 52.9 million in 2006, implying an average length of stay in the country of 2.5 nights. Foreign and domestic nights in total were up 2% to 351.2 million (foreign nights therefore accounted for 15% of total nights). Hotel occupancy in Frankfurt was down 2.5 percentage points to 62.2%, while Berlin enjoyed an increase in average occupancy of 6.3 points to 62.6%.

Although there were mixed performances from one market to another, both long- and short-haul sources recorded healthy growth. The largest recorded increase in arrivals was the 66% rise from Brazil, a nation of football fanatics. Arrivals from Iceland were up 53%, Portugal 44% and Ireland 34%. Among the more important European markets (ranked by size), arrivals from the UK were up 12%, the Netherlands 5%, France 8% and Poland 20%. More remarkably, there were increases of 9% from the USA, 16% from Canada and 6% from China.

Some of the growth rates for arrivals quoted by the German National Tourist Board (DZT) over the two months June and July were quite staggering: +110% from Portugal, +256% from Brazil, and +405% from Central America/Caribbean. In terms of increases in absolute volumes, the UK was the leading source, generating an additional 292,943 arrivals over the two months of the World Cup, while the USA, in second position, accounted for an additional 210,791 visitors. Research by DZT and IPK International suggested that 73% of all visitors over that period had travelled to Germany specifically for the World Cup.

The general impression seems to be that the FIFA World Cup represented a remarkable success for the German tourism industry in raising the image of the country as a friendly, interesting and attractive destination.

4.2 Domestic and Outbound Trends

According to IPK International's German Travel Monitor (Deutscher Reisemonitor), Germans made a total of 288 million trips in 2006, an increase of just 1% on 2005, of which 74% were domestic and 26% outbound. Since 2000 domestic trips have generally been increasing their share of the total market, and this trend continued in 2006.

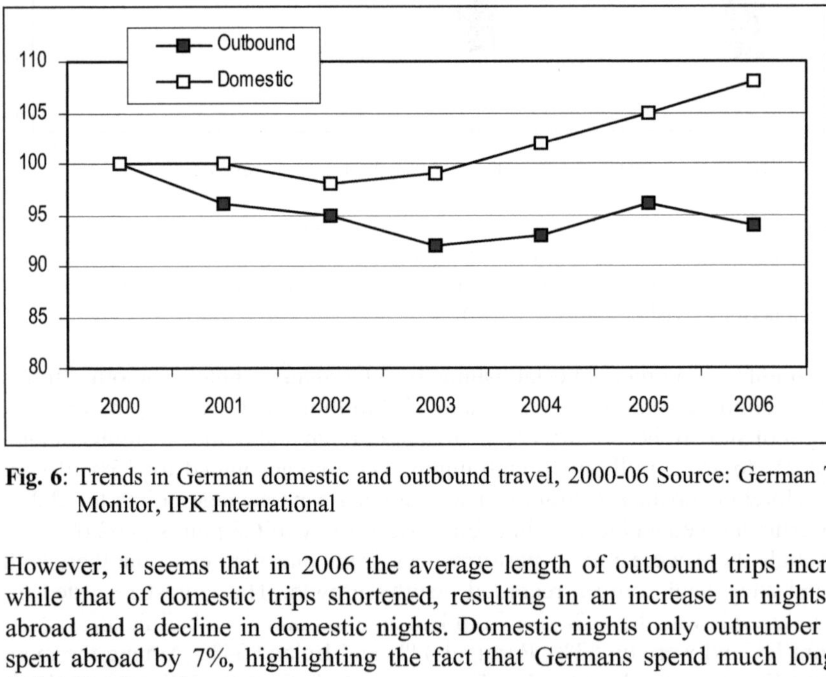

Fig. 6: Trends in German domestic and outbound travel, 2000-06 Source: German Travel-Monitor, IPK International

However, it seems that in 2006 the average length of outbound trips increased, while that of domestic trips shortened, resulting in an increase in nights spent abroad and a decline in domestic nights. Domestic nights only outnumber nights spent abroad by 7%, highlighting the fact that Germans spend much longer on their trips abroad.

There were only slight changes in spending at home and abroad in 2006, in euro terms. Total expenditure was flat, due to a 1% decline in spending on outbound trips and a 1% increase in domestic trip spend.

IPK International has been monitoring the German travel market since 1989, progressively increasing its scope and coverage. A total of 36,000 interviews are now conducted every year, making it the largest single travel survey in the country, with interviews carried out on a weekly basis.

Table 10: German domestic and outbound travel, 2006

	Total	% change on 2005
Trips (mn)		
Domestic	213	3
Outbound	75	-2
Total	288	1
Nights (mn)		
Domestic	738	-2
Outbound	692	3
Total	1,400	1
Spending (€ bn)		
Domestic	59	1
Abroad	63	-1
Total	122	0

Source: German Travel Monitor, IPK International

Domestic trips increased by 3% to 213 million in 2006. Within this total, holiday trips were up 2%, other private trips up 3% and business trips up 5%. A breakdown of the 90 million or so holiday trips shows that countryside recreation holidays increased by 7%, summer mountain holidays grew by 6%, trips to private events were up 6% and trips to public events up 43% – well above the average growth, and no doubt reflecting the FIFA world Cup once again.

Bavaria was once again the undisputed market leader in terms of domestic destinations visited by German holidaymakers in 2006, accounting for a 20% share of all domestic trips. In second position, some way behind, was Lower Saxony with 11%, followed closely by Baden Württemberg (10%) and North Rhine-Westphalia (9%). Germany's capital Berlin ranked eighth with a 7% share – lower than might be expected, given its increasing prominence as a world-class city, coupled with the rise in domestic short breaks by Germans.

Table 11: Leading destinations of German domestic holiday travellers, 2006

Rank	Destination	% market share
1	Bavaria	20
2	Lower Saxony	11
3	Baden Württemberg	10
4	North Rhine-Westphalia	9
5	Mecklenburg West Pomerania	8
6	Schleswig-Holstein	7
7	Saxony	7
8	Berlin	7

Source: German Travel Monitor, IPK International

The 2% decline in German outbound trip volume in 2006, which followed a promising upturn in the market in 2005, was entirely due to a drop in demand for foreign holidays (-3%). Other private trips, including VFR also fell by 3%, while business trips increased by 7% – confirming the recovery of this segment.

Table 12: German outbound travel by purpose of trip, 2006

Purpose	No. of trips (mn)	% market share	% change on 2005
Holidays	53	70	-3
Other private reasons	14	19	-3
Business	9	12	7

Source: German Travel Monitor, IPK International

The top seven destinations for German outbound travellers are all within Europe although, in contrast to long-haul destinations, a number of them lost market share in 2006. Italy, France and Turkey suffered heavy declines for the second year running – and Greece was supplanted by Switzerland in the top seven. Trips to Turkey fell by 13%, but this was after an 8% increase in 2005. Spain retained its position at the top of the league, but without showing much growth.

Table 13: Leading destinations of German outbound travellers, 2006

Rank	Destination	% market share	% change on 2005
1	Spain	17	1
2	Austria	15	-2
3	Italy	13	-6
4	France	7	-11
5	Turkey	6	-13
6	Netherlands	5	7
7	Switzerland	4	2
Top seven destinations		**67**	**2**

Source: European Travel Monitor, IPK International

The Netherlands showed the best growth of all, reversing previous year's decline, and Switzerland was also back in favour. The top seven destinations generated 67% of all German outbound trips, up 2% year on year.

The proportion of Germans who use the Internet to make their travel arrangements (46%) is closely comparable to the European average (see table 6), with rather more 'lookers' but fewer 'bookers'. It rose only slightly in 2006, by three percentage points, but this followed an exceptional increase in numbers of bookers in 2005.

One of the main drivers of online travel demand appears to be the expansion of German tour operators' dynamic packaging capabilities. Booking online is easy, quick and often a lot cheaper than buying traditional tour packages.

Table 14: German online travel, 2006

	% growth 2006/05	% market share 2006
Bookers	4	28
Lookers[a]	2	18
All online travel	3	46

[a] Use of the internet to research travel options, but not for booking
Source: German Travel Monitor, IPK International

Prospects for the German travel market are still uncertain. In 2006 there was a dramatic improvement in business confidence, but this was based on renewed competitiveness in export markets. And this competitiveness, in part, derives from measures which tend to unsettle consumers. So the private consumption was up only slightly in 2006.

In 2007 optimism has come back to German consumers. With nearly one million less unemployment in April 2007, compared to the previous year, the labour market is stabilizing faster than expected. Private consumption has meanwhile become the second base of a robust recovery. And wages seem to increase faster than the years before.

There is a clear sense that the economic prospects for laggards in the euro area, especially Germany, are improving. This is reinforced by the fact that the government has just raised its economic growth forecast for 2007 to 2.3% from 1.7% and is predicting an acceleration to 2.4% next year.

The GfK confidence index for May, based on a survey of about 2,000 people, rose to 5.5 from 4.4 in April. Economists were only expecting a reading of 4.7, the median of 28 forecasts, according to a Bloomberg News survey. GfK also believes that consumer sentiment will probably improve further. Confidence is increasing, the economy is doing well and the labour market is improving, which are all good indicators for rising consumption, including travel.

5 European Travel Trends in 2007

5.1 Overview of Main Trends in 2007

In 2006, European adults aged 15 years and over made 389 million trips abroad. Of these, 268 million (69%) were for holidays, 66 million (17%) for visits to friends and relatives (VFR) and other leisure purposes, and 54 million (14%) for business. Business travel, which had lost significant market share in recent years, has rebounded over the last couple of years, and the following table shows that the recovery was sustained in the first eight months of 2007, with business trip volume up a further 8%. It should be noted, nevertheless, that this estimate is based on trends in just 12 leading European markets (100,000 interviews) which, according to IPK International, account for roughly 57% of total European outbound trip volume.

Holiday travel, which stagnated in 2006, registered a 4% increase – thanks to strong growth from some of Europe's less traditional outbound markets – while, in contrast, VFR and other leisure trips (undertaken mainly for educational, medical and/or religious purposes) declined by 5% from January through August, after recording above average growth in 2006.

Table 15: European outbound travel, 2006-07

		% change 2006/05	% change Jan-Aug 2007/06[b]
Trips[a] (mn)	389	3	3
Short trips (1-3 nights long)	103	6	3
Long trips (4+ nights)	286	2	2
Holiday	268	3	4
VFR and other leisure	66	5	-5
Business	54	4	8
Overnights (mn)	3,700	5	0
Average length of stay (nights)	10	2	-4
Spending (€ bn)-	354	6	2
Spending per trip (€)	898	1	-2
Spending per night (€)	95	1	2

[a] Trips made by adults aged 15 years and over
[b] Based on trends in the first eight months of 2007 from the leading 12 source markets, which account for 65% of European outbound trip volume
Source: IPK International's European Travel Monitor

While sun & beach trips continue to dominate the outbound holiday market in Europe, they showed only modest growth in the first eight months of 2007 in terms of trip volume. The best growth sectors were city breaks and touring holidays, with mountain recreation (without snow) also attracting a healthy increase. However, poor snow conditions during the winter months at the beginning of 2007 resulted in a decline for winter sports holidays.

Total overnight volume, which reached 3.7 billion in 2006, stagnated from January through August this year, resulting in a lower average length of stay – a trend that has intensified in recent years as a result of the faster than average growth in short breaks using low-cost airlines.

Expenditure, meanwhile, rose by 2%, as did sending per night – because of the lower average stay – but spending per trip declined by 2%.

Among the trends confirmed by the European Travel Monitor in the first eight months of 2007 was a faster increase in long-haul travel (4%) than in short-haul trips (2%) – a trend that had been forecastrd by the Pisa Forum in 2006.

Thanks to the increasing weakness of the US dollar against the euro and sterling, travel to the USA picked up quite strongly from January through August, and sub-Saharan destinations have performed well in the European market. According to IPK the sharpest rises were for Asia, especially China and Indochina.

Travel to North Africa by Europeans has shown above average growth, and there have been healthy rises for the southeast and southwest Mediterranean countries, but eastern, western and northern European countries have attracted lower than average growth from European markets.

Despite poor snow conditions during the winter season 2006/07, renewed terrorist attacks (eg in the UK), floods in western Europe (UK) and forest fires in southern Europe (Greece and Italy) – not to mention a lack of mega sporting or cultural events to stimulate tourism demand – the European market confirmed its resilience.

5.2 Leading Markets

Europe's two leading travel source markets, Germany and the UK – which, accounted for 35% of total European outbound trip volume – have turned in disappointing performances this year so far, and are not expected to show much improvement before the end of 2007. In Germany's case, the market's stagnating volume was attributed by Pisa Forum participants to uncertainties over employment prospects, together with the negative impact of the three percentage point rise in VAT at the start of the year.

The decline in the UK market, on the other hand, is blamed on different factors – the government's decision to sharply increase the Air Passenger Duty, as well as a possible saturation in demand for secondary short breaks using low-cost carriers, and increased hassles associated with travel through UK airports as a result of stepped-up security and immigration checks. However, as representatives of the USA, Canada and Kenya in Pisa confirmed, while total trip volume has stagnated, demand from the UK for long-haul destinations has recorded a healthy increase.

Meanwhile, in terms of percentage growth over the first eight months of 2007, the best performers were either those markets more traditionally considered as destinations or domestic markets – Spain and Italy, for example – or emerging markets such as Russia. There were also above average increases for Ireland, Scandinavia and France.

Table 16: Leading European outbound travel markets, January through August 2007

Market	% growth in trips
Spain	11
Russia	10
Italy	7
Norway	5
Ireland	5
Sweden	5
France	5
Finland	2
Belgium	1
Denmark	0
Germany	0
UK	-2

Source: IPK International's European Travel Monitor

In the case of France, and probably also Spain and Italy, outbound travel demand in recent years has been boosted by the introduction of the euro, since it has made prices more transparent across Europe and drawn attention to the fact that foreign travel can be less expensive than holidaying at home.

It should be noted, of course, that the preliminary 2007 results are based on the top 12 markets in Europe only, so the higher growth expected from some of the region's smallest members is not yet evident and will only be confirmed after the year has ended.

5.3 Destination Trends

Trends in European outbound travel this year have again been influenced by a number of factors. On the positive side, the strong (and appreciating) euro, the continuing healthy economic situation in many countries, the expansion of low-cost/no-frills flights, and the poor weather during much of the summer in Northern Europe, all stimulated demand for outbound travel in the first eight months of the year.

Long-haul destinations make up most of the 'superstars' – destinations attracting more than 10% growth out of the European market – as the following table shows, although Morocco, Serbia, Montenegro and Iceland all feature among the winners.

Table 17: Performance of different destinations in the European market, January through August 2006

>10%			
China	Japan	Cambodia	Vietnam
Kenya	Morocco		
Serbia	Montenegro	Iceland	
5-10%			
Belgium	Finland	Germany	Portugal
Croatia	Latvia	Lithuania	Poland
Malta	Egypt	Middle East	South Africa
India	Malaysia	Thailand	Switzerland
Brazil	USA		Turkey
1-4%			
Denmark	France	Ireland	Czech Republic
Netherlands	Spain	UK	Bulgaria
Tunisia	Italy	Greece	
Canada	Dominican Republic		
Australia	New Zealand		
Weak performance (0% and below)			
Austria	Cyprus	Hungary	Russia
Singapore	Sri Lanka	Israel	Estonia
Cuba			

Source: IPK International's European Travel Monitor

5.4 Transport

For the first time for several years, there was an increase in travel by rail on European outbound trips. While the total volume of rail-based trips remains small, it was a double-digit increase, attributed in part to the increased hassles of air travel due to stepped up security and immigration controls at airports.

Although some delegates at the Pisa Forum questioned whether Europeans are perhaps starting to avoid air travel as a way of reducing their carbon emissions, there is no clear evidence of such a trend.

Meanwhile, airline trips increased by 5%, due mainly to the continuing boom in demand for low-cost/low-fare airline travel, more usually dubbed 'no-frills flights'. According to the European Travel Monitor, trips involving flights costing less than €150 return increased by 17% in volume in the first eight months of this year, with the result that they now account for more than 39% of total trips abroad by air made by Europeans.

It should be noted that trips involving airline travel account for well over half of all outbound trips made by Europeans, although this average clearly masks wide variations from one source country to another.

5.5 Travel Distribution

Use of the Internet for online booking as opposed to simply 'looking' – gathering information prior to booking a trip – continues to grow very fast in Europe. As a result of a 13% increase in online bookings (for at least part of a trip) during the first eight months of 2007, their share has now reached 36% of total trips abroad. The equivalent share was just 19% in 2003. Moreover, in some markets the share is over 65%.

If those using the Internet to research their foreign trips – e.g. the online 'lookers' – are included in the count, the share of Internet users in Europe for travel is now 50%. Clearly, the incidence of Internet bookings among the smaller, less mature European markets is lower, but it is growing fast – much faster, in fact, than online travel 'looking'.

Table 18: European online travel trends, January through August 2007

	Market share Jan-Aug 2007	% change Jan-Aug 2007/06
Online booking	36	13
Online 'looking'	14	3
All internet users	50	10
Non-internet users	50	-8

Source: IPK International's European Travel Monitor

The Internet is now nearly twice as important generally as travel agents as an information source, although the travel trade is still very important in terms of travel distribution. And, given the rise in dynamic packaging offered directly by tour operators to clients booking online, the travel trade is not about to lose its importance as a distribution channel.

Table 19: Information sources used by European outbound travellers, January through August 2007

Source	% sharea
Internet	47
Travel agency	24
Friends/relatives	15
Travel guide	7
Travel brochure	7
Newspaper	2
Tourist office	2
TV	1
No response	7
Others	16

[a] Multiple responses possible
Source: IPK International's European Travel Monitor

While more than one third of all bookings are made at least partly online, 25% involve travel agents, 9% are booked direct with hotels and 7% direct with transport companies. Just over 20% of all European outbound trips do not involve any advance bookings at all.

Table 20: Europeans' travel booking channels, January through August 2007

	% of trips[a]
With help of internet	36
Travel agents	25
Direct with hotels	9
Direct with transport companies	7
Other[b]	14
No advance booking	26

[a] Multiple responses possible. In particular, bookings through travel agents, hotels, transport companies and other channels may also be made with the help of the internet
[b] e.g. Newspapers, clubs, staff associations
Source: IPK International's European Travel Monitor

Contrary to popular opinion, package holidays are not out of favour among Europeans. In the first eight months of 2007, as many as 59% of all trips booked from January through August 2007 were for packages (i.e. transport plus hotel) – an increase of 8% over the same period in 2006. A significant share of these packages were, however, self-tailored online, primarily as dynamic packages.

Bookings for transport only (a 24% share of total trips) rose by just 2% and accommodation-only sales were up 11%, accounting for a 16% share.

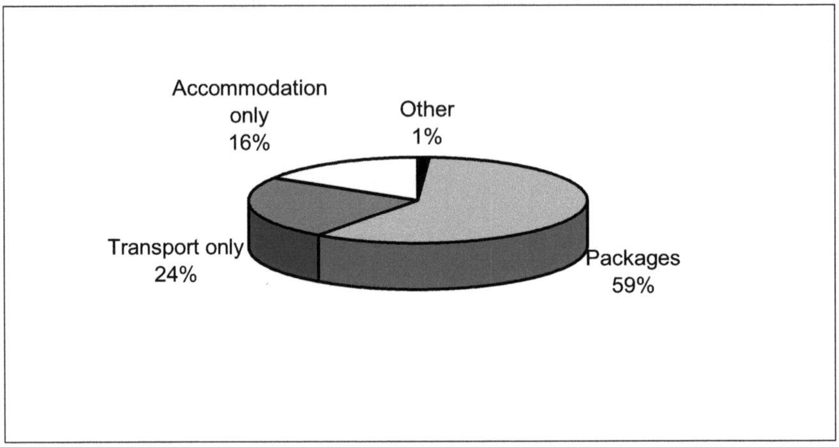

Fig. 7: Breakdown of European pre-bookings, Jan-Aug 2007;
Source: IPK International's European Travel Monitor

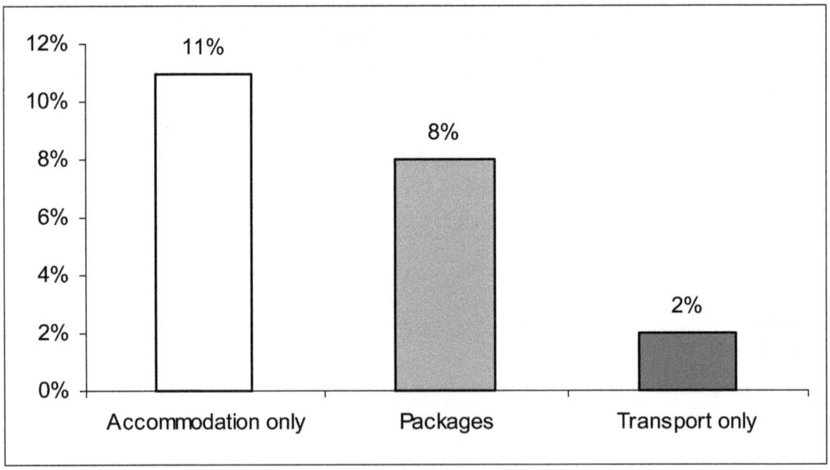

Fig. 8: Annual growth of European pre-bookings Jan-Aug 2007 vs. Jan-Aug 2006;
Source: IPK International's European Travel Monitor

Climate Change and Its Impact on the Tourism Industry

Roland Conrady, Stephan Bakan

1 Introduction

Nowadays it is common knowledge among climate specialists that an anthropogenic climate change is taking place with a significant influence of human activities on the global climate. Although "climate skeptics" are still not convinced of the significance of the arguments presented by climate research, they usually argue without satisfactory scientific knowledge and use populist conclusions.

Since approximately the end of 2006, the topic climate change has been a key issue. In fact, the fourth Assessment Report, published in February 2007 by the Intergovernmental Panel on Climate Change (IPCC), contributed significantly to centering people's attention on this issue.[1] Another impulse for this topic was given by the Stern Review on the Economics of Climate Change, commissioned by the British Government. Sir Nicholas Stern calculated, for the first time ever, the cost climate change would have on the national economy, and thrust figures amounting to 20% of the GDP into the full glare of the public eye throughout the world.[2] Furthermore, one of the most successful documentaries of all times – *An Inconvenient Truth* by Nobel Peace Prize Laureate Al Gore – attracted the attention of a larger audience. Now, conferences on climate change are taking place almost every month. Indeed, climate change has become a key topic on a global political level (i.e. the G8 Summit in Heiligendamm/Germany 2007) as well as at a national political level.

Even the tourism industry has begun to show a highly visible interest in climate change. The climate conference, held in Djerba/Tunesia by the World Tourism Organisation (UNWTO) in April 2003 and the ensuing signed Djerba Declaration have hardly been noticed in the global tourism industry. In contrast, the International Conference on Climate Change and Tourism, held in Davos/Switzerland and hosted by the UNWTO in October 2007, attracted considerable attention. In Germany, climate change was one of the key topics at the Tourism Summit 2007

[1] See IPCC (2007)
[2] See Stern (2006)

hosted by the Federal Association of the German Tourism Industry as well as at the World Travel Market 2007 in London. The International Tourism Exchange ITB Berlin can take credit for sensitizing the industry to climate change and tourism at the ITB Convention 2007.

2 Observed Phenomena Due to Climate Change

Nowadays, the following perceptions on climate change are considered quite accurate:[3]

- The concentration of carbon dioxide in the atmosphere has significantly increased since approx. 1850, from the typical value of 280 ppm[4] in the warm periods for at least 400,000 years to the current value of 380 ppm.
- Mainly human beings are responsible for this increased level ("anthropogenic climate change"), especially as a result of burning fossil fuels (crude oil, natural gas and coal), but also through deforestation. A small fraction of the augmentation can be traced back to natural causes such as fluctuations in solar activity.
- Carbon dioxide (CO2) is – like methane (CH4) and laughing gas (N2O) – a greenhouse gas, or in other words, a gas which affects the climate and changes the Earth's radiation budget[5]: An increase in its concentration leads to a rise in the surface temperature. If the concentration doubled, the average global warming effect would most likely be 3 +/– 1 °C.
- The temperature increased considerably in the 20th century: approx. 0.6 °C globally and 1.0 °C in Germany. Furthermore, the temperatures over the last ten years have been the warmest since the beginning of records in the 19th century and most likely for at least several hundred years beforehand.
- Since 1978, the surface area of the perennial Artic ice fields has decreased by 7% and the winter ice field expansion by 3% every decade. According to the latest findings, the ice in the North Pole region has diminished by one million square kilometers and is currently only 3 million square kilometers. In fact, there is no ice in the North West Passage[6]
- The mean sea level has risen by an average of 3 mm per year since 1992, whereas throughout the whole 20th century it rose on the average only 1.5 mm per year.
- The precipitation structure has also changed: the subtropical regions are becoming even dryer and the northern regions more humid.

[3] See Rahmstorf, Schellnhuber (2006) p. 52 f.

[4] ppm = parts per million

[5] Greenhouse gases in the atmosphere prevent the reflection of solar energy and thus increase the temperature of the atmosphere.

[6] The North West Passage runs from the Bering Strait along Canada's northern shore and Greenland's West coast.

The last time the global temperature rose at such a strong rate was about 15,000 years ago, when the last ice-age came to an end: In that period, the Earth's temperature rose approx. 5 °C. But this global warming took place over a period of 5,000 years, whereas humankind may now face a similar dramatic climate change within a century.

3 Climate Projection

The climate changes that have been visible so far are only small precursors to much larger changes, which may come to pass if the concentration of greenhouse gas continues to increase. It has to be taken into consideration that every additional gram of CO_2 remains in the atmosphere for approx. 100 years. In other words, even if global CO_2 emissions were cut back strongly, a reduction in the CO_2 concentration in the atmosphere would not be noticeable in the short term.

Along with the analysis of the observed climate changes, existing climate models are used to estimate future changes. In this method, estimated societal developments and emissions are presented as "scenarios". The various scenarios for future emissions contain different assumptions about humankind's economic, technical and sustainability development and vary from "just like always" to "drastic measures".

Using plausible scenarios and taking remaining variables for the calculability of the climate system into account, the IPCC determined in its latest report that a global temperature increase of 1.4 and 5.8 °C is to be expected by the year 2100 (above the 1990 value). However, according to more recent studies, higher values are not to be disregarded, should the feedback in the carbon cycle increase. On a side note, a mean global temperature increase of 6 °C is the estimated temperature difference between an ice age and an age with excessive heat.

Figure 1 demonstrates the *temperature increase* within the three scenarios selected by the IPCC in respect to different regions. In general, apart from the Polar Regions, the calculated temperature increase is noticeably higher in the interior than over oceans and coastal regions. On the whole, the temperature increase is higher in winter than in summer.

In the best case scenario B1, which incorporates "drastic measures" to avoid climate change, the mean global temperature will increase by 2 degrees and stabilize on a long-term basis. Many ecologists feel that, for the most part, the global ecosystem is just up to tolerating this increase.

One of the consequences of global warming is that the *glaciers and polar caps melt* and the *permafrost soil thaws*.[7] The first one threatens ecosystems and contributes to the rising sea level; the latter damages the infrastructure, i.e. transport routes and buildings. and releases vast quantities of more methane which, as a

[7] Polar bears, whose very basis of existence is threatened by the diminishing ice floe, have become symbols of an endangered fauna.

greenhouse gas, is still far more damaging to the global ecosystem than carbon dioxide.

Due to the increased amount of melt water and to an even greater degree of thermal expansion in the sea, an *increase in the sea level* of about ten centimeters is to be expected before the end of the century. As a result, a large part of the coasts and beaches would be lost. For every millimeter the sea level rises, a 1.5 m strip of beach is lost.

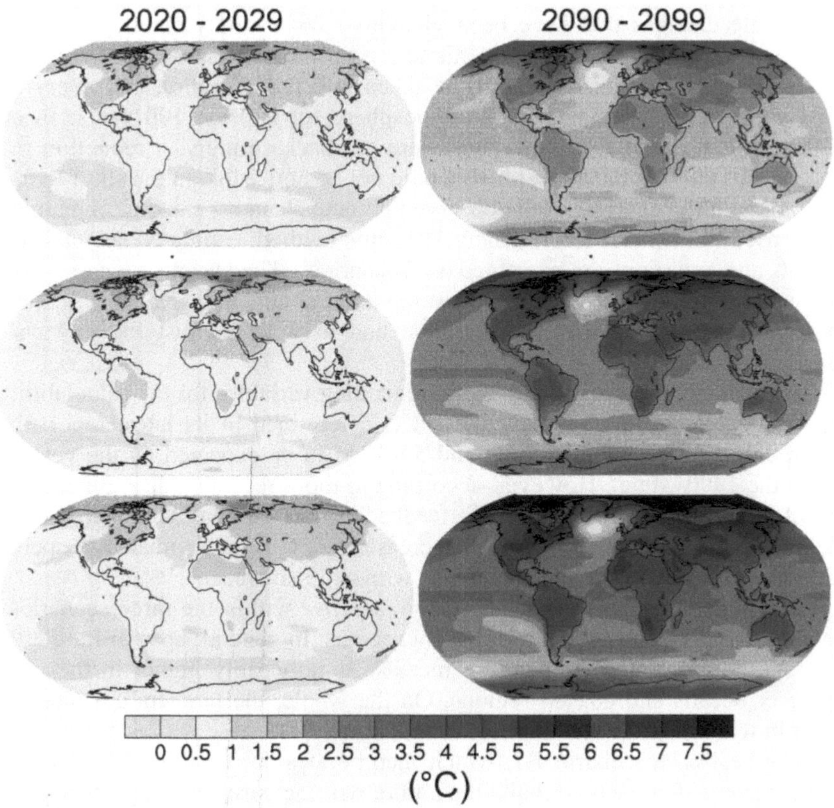

Fig. 1: Estimated Temperature Changes by 2100[8]
Note: The two top illustrations demonstrate the temperature development as per scenario B1 ("best case"). The two middle illustrations represent scenario A1B ("real case") and the bottom illustrations depict scenario A2 ("worst case").

[8] IPCC (2007)

The water circulation is becoming stronger. Consequently, the subtropical regions experience less *rainfall* and the regions in the mid and polar latitudes experience heavier. Central and Northern Europe are expected to have less rainfall in summer and more in winter. Even the number and intensity of *tropical storms* i.e. hurricanes is expected to increase.

4 Climate Change and Tourism

As in most other industries, the travel and tourism industry has an ambivalent relationship with climate change: The travel and tourism industry produces ("culprit") and suffers from ("victim") climate change at the same time.[9]

The travel and tourism industry is a producer, emitting greenhouse gases throughout all levels in the tourism value chain. The first culprit of course is transport. Tourism without mobility, or without transport mechanisms, is unimaginable, and transport is almost always interlinked with CO_2 emissions.[10] Transport contributes to approximately 25% of the CO_2 emissions and is continually increasing; transportation has the highest growth rate among CO_2 producing sectors.[11] In addition, the energy consumption in vacation destinations for heating or air conditioning or for operating specific attractions (i.e. Ski Dubai, Center Parcs or Tropical Island) is quite high.

The tourism industry suffers in all of the three elements of the tourism products: *Attractions, Amenities* and *Access* (the three As).[12]

Attractions will lose their attractiveness caused by climate changes. The attraction to ski resorts will diminish due to unreliable snow conditions. In the Alps for example, snow conditions are only reliable above 1200 m. If the temperature rises by 2 °C, reliable snow conditions will not be found until 1500 m. In Switzerland, nowadays 85 % of the ski resorts have reliable snow conditions; in the future this number will decrease to 44 %. From the investors' perspective, amortizing investments in winter sports facilities will become more difficult. Alpine ski resorts need to be open for business 70 days each year in order to be profitable. It is already difficult to assume such consistency. In contrast, North American ski re-

---

[9] In certain aspects, the tourism industry could feel like winners as a result of the climate change, i.e. along the North and Baltic Sea Coasts.

[10] Exceptions include niches market such as cycling tourism. See Becken/Patterson (2006) for the methodical approaches to measuring to what extent the tourism industry influences carbon dioxide production

[11] The airline industry has been severely criticized for this aspect since approximately one year ago. For some reason, airlines are reputed to be the No 1 climate killer. In fact, the airline industry only accounts for 2–3% of the global CO_2 emissions. In comparison, the shipping industry generates 4% of the global CO_2 emissions.

[12] Attractions are natural and cultural features such as a beautiful landscape or architectural monuments. Amenities are tourism facilities which include in particular hotels. Access denotes streets, airports, flight and rail connections.

gions do not seem to have been affected. Studies on ski tourism in North America could not identify any significant effects resulting from climate change.[13]

In certain summer destinations, i.e. Spain, the increasing heat is making the traditional vacationing months of summer unattractive. If the global temperature increased by a mere 2 °C, summers would gain 21 more days of extreme heat.

As a result of the rising sea level, coastline loss is sure to be bemoaned. It is estimated that an over 108 m stretch of beaches along the coastline will be lost by the year 2030. "This would wipe out beaches across the globe and coastal amenities such as hotels, golf courses and retail facilities would be threatened" (McGuire 2006).

Flat atolls, such as the Maldives in particular, are endangered. Realistically seen, the extremely popular Maldives archipelago will most likely disappear under the ocean.

Since natural attractions, such as coral reefs, are very sensitive to any increases in temperature they are also threatened by global warming. Cultural attractions, such as monuments are also faced with destruction (i.e. Venice).[14]

Although some destinations lose tourism appeal, others will presumably gain attractiveness. This is the case especially in the cooler northern countries (Scandinavia, Canada, Germany etc.). Due to the rising temperature, Germany will become more attractive as a tourist destination. The Potsdam Institute for Climate Impact Research has calculated that the number of potential days for bathing in Baden-Wuerttemberg will increase by 4% within thirty years[15].

Amenities are threatened by extreme weather conditions. An increase in tornadoes, hurricanes and forest fires will endanger tourism facilities and a rise in the sea level will jeopardize resorts located near the coast. Water shortage will also negatively affect the attractiveness of amenities.[16] However, water shortages are less likely to be of concern in regions with lots of water (i.e. Northern Europe), whereas the situation will intensify in regions that already suffer from water shortage (i.e. regions in Asia and Africa).

In a study commissioned by Halifax Travel Insurance, the Benfield Hazards Research Center addressed the impacts of climate change in individual regions. Table 1 illustrates the essential results.[17]

Access becomes difficult due to the increase in costs in the passenger transport sector.[18] Traveling with airplanes, automobiles, trains and ships will become more expensive. The increasing shortage of fossil fuels in respect to consumer demand drives costs higher. Hence, the price of fuel can also be expected to rise. Mechanisms for pricing CO_2 emissions will also affect the cost increase for transport. Cost savings brought on by introducing alternative fuels are not to be expected be-

[13] Compare with Scott et al. (2006)
[14] Refer to the European Union's research project Noah's Ark: http://noahsark.isac.cnr.it/about.php
[15] See Stock (2005) p. 117
[16] Golf resorts and swimming pools depend on sufficient water supplies.
[17] Consult McGuire (2006)
[18] Cargo transport will of course also become more expensive.

cause the high development costs would initially lead to a rise in price for such innovative transport methods.

From the political point of view, increasing the cost of mobility is beneficial for dampening the demand for transport. Due to the disproportionate growth in the air transport industry, political parties have pinpointed it in particular as a key player in climate change and have instigated pricing measures for air pollutant emissions in air transport.

"In December 2006, the European Commission published a proposal for a directive on the inclusion of aviation into the European Union's emission trading scheme for stationary sources. According to the proposal, aircraft operators will be obliged to hold CO_2 emission allowances for all intra-European flights from January 2011 onwards. In 2012, the geographical scope is supposed to be extended to include all flights arriving or departing from airports within the EU."[19] Air transport's inclusion in emission trading has not been without controversy and has not yet been decided up to now, but is nevertheless to be expected. The opposition asserts that such an action will be detrimental for non-European airlines, which will not receive equal treatment: A distortion in competition unfavorable for European airlines is not acceptable.[20] Furthermore, airlines criticize the fact that the additional cost burden, which results from acquiring emission certificates, will make investment potential in environmentally friendly technology more difficult.

Transferring the cost increase for emission trading will most likely cause the ticket prices to rise by 1–4 %. Table 2 details possible effects on consumer demand as a result of the cost increase, calculations performed at the German Aerospace Center.[21]

All in all, conclusion of the European Travel Commission can be agreed with. The consequences that the climate change will have on the global travel and tourism industry are going to be extreme: "In less than 25 years, climate change will have a radical impact on the global travel industry."[22]

[19] Scheelhaase et al. (2007) p. 1

[20] In long distance transport, European airlines would be at a disadvantage, if they were for example to have to swallow cost increases for passenger transport to India whereas their competitors such as Emirates could transport passengers to India via the hub in Dubai with no additional costs involved.

[21] Scheelhaase et al. (2007) p. 5; Lewis et al. (2007)

[22] Franklin 2006

Table 1: Main destination impacts influencing tourism[23]

European summer destinations		**Summer destinations in the rest of the world**		**Winter sports destinations**
Southern Spain (Costas; Balearics):	Higher maximum temperatures and heat waves Water shortages, Increased wildfire risk, Potential for malaria and other vector-borne diseases, Beach erosion, Increasing aridity and desertification,	United States (Florida):	More powerful hurricanes, Coastal inundation and flooding due to storm surges, Erosion and loss of beaches, Damage and loss of coral reefs and other coastal ecosystems, Wildfires, Emergence of vector-borne diseases such as malaria and dengue fever	Rise of snowline by up to 300 m in Europe, Less snow cover, Shorter season, Increased avalanche risk, Increased potential for landslides
Canary Islands:	Potential for more powerful storms, Beach and cliff erosion	The Caribbean Islands:	More powerful hurricanes, Erosion and loss of beaches, Water and power shortages, Damage and loss of coral reefs, Emergence of vector-borne diseases	
France (Provence and the Côte d'Azur):	Higher maximum temperatures and heat waves Increased wildfire risk, Water shortages, Flash floods, Beach erosion and damage to wetlands	Australia (Queensland):	Decreased cloud cover and consequent increased ultraviolet radiation levels, Potential for emergence of vector-borne diseases, More powerful cyclones, Damage and loss of the Great Barrier Reef	

[23] Compare to McGuire (2006)

Greece (the islands):	Higher maximum temperatures, Water shortages, Increasing aridity and desertification	North Africa (Morocco and Tunisia):	Higher maximum temperatures, Beach erosion, Potential for more powerful storms on Morocco's Atlantic coast, Dust storms, Desertification
Italy (Amalfi Coast and Tuscany):	Higher maximum temperatures and high humidities on Amalfi coast, More heat waves in Tuscany, Flash floods, Beach and cliff erosion	India and the Indian Ocean (Goa, Kerala, Maldives, Seychelles):	Potential for more powerful cyclones in Goa and Kerala, Erosion and loss of beaches, Inundation and flooding of coastal zone and further inland, Damage and loss of coral reefs in Seychelles and Maldives, Potential evacuation of some islands in Maldives due to saltwater penetration of aquifers.
Malta and Cyprus:	Higher maximum temperatures, More extremely hot days and hot nights in Malta, Water shortages		

Table 2: Potential demand impacts selected flights

Airline	Ryanair	Lufthansa	Lufthansa	Emirates
Routing	Frankfurt Hahn – London Stansted	Frankfurt – London Heathrow	Frankfurt – Singapore	Frankfurt – Dubai
Absolute fare increase due to EU-ETS[a]	1,14 €	1,57 €	20,06 €	8,20 €
Average fare per passenger and segment	44,00 €	136,00 €	602,00 €	233,75 €
Relative fare increase	2,57 %	1,15 %	3,33 %	3,51 %
Demand reduction when costs are shifted fully and equally to all passengers	- 3,41 %	- 1,28 %	- 2,17 %	- 2,29 %

[a] ETS = Emission trading scheme

5 Preventive and Adaptation Strategies in the Travel and Tourism Industry

The cause and effect relationship *climate change impacts tourism* implies the importance of adaptation strategies. The inverse cause and effect relationship *tourism impacts climate change* enforces the need to develop preventative strategies.[24]

To begin with, possible methods to develop preventive and adaptation strategies in the travel patterns have to be analyzed.

The most preventative method would without a doubt be to abstain from travelling, and thus also abstain from environmentally harmful transport activities.

This option is a theoretical case. In practice, people are hardly inclined (in the case of private or holiday trips) or in the position (in the case of business trips) to refrain from travelling. Empirical studies also confirm that strong tourism drivers, to be precise population and income development, are much more predisposed to increase the number of tourists than dampen climatic impacts.[25] Whether tourists are willing to compensate fewer trips with longer stay is questionable.

If anything, tourists will probably adapt to the climate change by altering their travel patterns. Travellers are not indifferent to the climate at their place of residence or the destination. Empirical studies have confirmed that for example 80%

[24] For the research agenda for the two cause and effect relationships, refer to the articles in the Journal of Sustainable Tourism, volume 14, N° 4, 2006.

[25] Compare with Hamilton et al. (2006) p. 263

of Brits claim that better weather is a factor for vacationing outside of Britain. German travellers, on the other hand, rank weather as the third most important criterion for making tourism purchasing decisions.[26]

Price increase in the transport sector could cause tourists to make fewer long distance trips and search for destinations near to home.[27] Such a variation in behavior would shift the tourism flows away from the poorer countries and in favor of the wealthy countries. Africa, for example, is only accessible for most tourists by means of long distance flights. Thus, development and political climate objectives blatantly clash.

Changes in tourism seasonality are quite plausible. Tourists from Northern and Central Europe, for instance, will visit Mediterranean destinations less frequently during the summer months, since it will be too hot. Trips to the Mediterranean will be much more comfortable in winter or spring. In an empirical survey among Germans, 85.2 % of the people surveyed stated that they would plan their vacation in a different season if the temperatures became more pleasant.[28] Consequently, the tourism season would be prolonged and the normally strong seasonal fluctuations in tourist arrivals and overnight stays could stabilize to a certain extent (destinations such as Dubai in the United Arab Emirates already exhibit this pattern). In the end however, a decline or at the best a stagnation in tourist figures is to be expected: Tourists lost in summer months cannot be recuperated in winter and spring.[29] Whatever the case may be, the Mediterranean destinations should definitely consider catering to travel motives other than "sun, sand and sea" in the future.

Another plausible outcome includes redirecting the global tourism flows. In Europe, a clear north–south pattern has dominated the travel flow for many years: Northern and Central Europeans travel south. In the future, Northern and Central Europeans will select domestic regions because temperatures will become more agreeable and southern destinations too hot. Conversely, Southern Europeans will head north seeking regions with more bearable climates. Hence, the traditional north–south migration will transform into a south–north migration.[30]

In the Alps, tourists will prefer to head higher to reach ski resorts with reliable snow conditions. Ski resorts in lower altitudes will have to cater for different non-ski travel needs.[31] Alternatives include culture, wellness, events etc. An adaptation strategy could be to produce artificial snow, exchange ski lifts for chair lifts, ski run management and opening up new ski runs at higher altitudes.

---

[26] Weather and climate are of course not to be mistaken for one and the same. Nonetheless, it may be assumed that the empirical findings on the impacts of weather can, with certain limitations, be transferred to information on climate change impacts.

[27] Consult the results from the simulation model proposed in Hamilton et al. (2006).

[28] Compare the studies presented by the International Tourism Exchange ITB Berlin and the travelchannel.de, in which in spring 2007 a total of 1,200 people were surveyed about how climate change would impact their travel behavior.

[29] Empirical study by Amelung and Viner (2006); Perry (2006); Hamilton et al. (2005)

[30] See Bigano et al (2006)

[31] Destination such as Kitzbühel in Austria will be under greater pressure to adapt. Kitzbühel lies at an altitude of only 760m.

In coastal regions, the rising sea level will present many challenges. As part of a research project for protecting the German North Sea Coast, researchers have proposed three adaptation strategies, which can even be combined. The first requires increasing the height of the existing dikes by another 70 cm, the second considers erecting a second dike inside the first dike line and the third option is building a flood barrier along the mouth of rivers and bays.

The global air transport industry is already embarking on extensive preventative strategies.

In the area of aircraft construction, clear endeavors are already underway – including even a first time cooperation between Airbus and Boeing – to find methods to raise fuel efficiency. By the year 2020, Airbus has promised to reduce its CO_2 emissions in new aircraft models by 50 % over the 2000 values, and should even be able to reduce nitrous oxide by 80 %.

Saving potentials for kerosene consumption can be achieved if the following actions are taken:[32]

- Usage of propane and propeller motors: 11–15 %
- Shorter air routes: 12 %
- Optimized turbo jet engines: 12 %
- Improved wing form: 7 %
- Increased use of light-weight construction material: 5–6 %
- Shifting the motor to the tail-end: 5 %
- Installing a gear box: 4–5 %
- Exchanging hydraulic systems for electric systems: 3 %
- Decentralized electric power supply with mini fuel cells: 3 %
- Use alternative fuel: up to 3 %
- New surfaces fuselage (shark skin): 1–2 %

"There are a number of alternative fuel options for aviation. Synthetic liquid fuel or synfuel is any liquid fuel obtained from coal, natural gas or biomass. Bio-jet fuel is made from converted agricultural oil crops like soya.[33] Ethanol fuel can be combined with gasoline in any concentration up to pure ethanol (E100). Hydrogen is being explored for application in fuel cells and might become an option for aircraft engines from 2050. In mid-term hydrogen may be relevant for Auxiliary Power Units (APUs), ram air turbine (RAT) and distributed power units. These applications will generate large fuel savings on the ground, lower noise and lower NOx."[34] Currently, research on using solar energy in aircrafts is in an experimental stage.[35]

[32] Kiani-Kress (2007) p. 95. See www.enviro.aero.

[33] When Sir Richard Branson, the founder of Virgin, began producing biofuel (*Virgin Fuels*) on the major industrial markets, the airline industry took full notice.

[34] Air Transport Action Group (2007); consult the International Air Transport Association (2007) in www.iata.org.

[35] www.solarimpulse.com.

Optimizing air traffic management is also quite effective. In Europe, the Single European Sky is to be put into practice. The goal is to optimize air routes and avoid needless holding patterns. This plan could lead to kerosene savings in the amount of 8–12 %. Lufthansa German Airlines alone could save approximately 142,000 tons of kerosene per year, which corresponds to eleven daily flights between New York and Frankfurt. Moreover, another expected development in the airline industry is to differentiate takeoff and landing fees at airports according to CO_2 emissions. This strategy would make it more sensible to use environmentally friendly aircraft.

Small contributions to climate protection are also to be expected from public projects such as the emissions calculator at Atmosfair. On this website, end consumers can calculate the pollutant emissions for a selected flight and have the opportunity to finance reforestation projects with a donation. The donation is more likely to salve consumers' conscience than help the climate.[36]

6 Key Proposals for Designing Future Climate Protection Strategies

Political decision makers should develop incentives, which could stimulate both producers and consumers to take an environmentally friendly stance, without having to economically weaken the travel and tourism industry. It is also extremely advisable to provide travellers with useful information about environmentally harmful or environmentally friendly travel. Currently, environmentally friendly behavior is problematic since many travellers have no idea about the actual impact their travel patterns have on the climate. On the other hand, they lack information about alternative environmentally friendly travel products on the market which they could take advantage of to change their travel behavior. Labeling products with an accredited certification could be a first step on a long road to solving problems.

[36] www.atmosfair.de

References

Amelung B, Viner D (2006) Mediterranean Tourism: Exploring the Future with the Tourism Climatic Index. Journal of Sustainable Tourism. 14:349–366

Air Transport Action Group (2007) In: www.atag.org

Becken S, Patterson M (2006) Measuring National Carbon Dioxide Emissions from Tourism as a Key Step Towards Achieving Sustainable Tourism. Journal of Sustainable Tourism. 14:323–338

Bigano A, Hamilton JM, Tol RSJ (2006) The Impact of Climate Change on Domestic and International Tourism: A Simulation Study. In: http://www.feem.it/Feem/Pub/Publications/WPapers/default.htm

Franklin, RK (2006) Changing with the Climate. Presentation of the European Travel Commission at the Pisa World Travel Monitor Meeting of IPK International, Nov. 1st 2006, Pisa/Italy.

Hamilton JM, Maddison DJ, Tol RSJ (2005) Climate change and international tourism: A simulation Study. Global Environmental Change. 15:253–266

International Air Transport Association (2007) www.iata.org.

IPCC (2007) Climate Change 2007: The Physical Science Basis, Summary for Policy Makers. Contribution of Working Group I to the Fourth Assessment Report of the Intergovernmental Panel on Climate Change. Paris, February 2007

ITB Berlin and travelchannel.de (2007) Empirical Study about travel intention due to climate, 1.200 test person in an online survey. Unpublished Study. Berlin 2006.

Kiani-Kress R (2007) Neue Wege. Wirtschaftswoche Nr. 25 vom 18.06.2007. p. 95

Lewis D, Ringbeck J, Salomon R. Defining Incentives to Cut Carbon Emissions in the Aviation Sector – Inevitable Steps to Fight Climate Change? Governors Meeting for Aviation, Travel and Tourism, World Economic Forum Annual Meeting 2007.

McGuire B (2006) Holiday 2030 – Report produced for Halifax Travel Insurance, in: http://www.benfieldhrc.org

Perry A (2006) Will Predicted Climate Change Compromise the Sustainability of Mediterranean Tourism? Journal of Sustainable Tourism. 14:367–375

Rahmstorf S, Schellnhuber HJ (2006) Der Klimawandel. Diagnose, Prognose, Therapie. 2. Aufl. München 2006

Scheelhaase J, Grimme W, Schaefer M (2007) European Commission Plans Emissions Trading for Aviation Industry. Aerlines e-zine edition, Issue 36

Scott D, McBoyle G, Minogue A (2006) Climate Change and the Sustainability of Ski-based Tourism in Eastern North America: A Reassessment. Journal of Sustainable Tourism 14:376–398

Stern N (2006) Stern Review Report on the Economics of Climate Change www.hm-treasury.gov.uk.

Stock M (2005) KLARA: Klimawandel – Auswirkungen, Risiken, Anpassung. PIK-Report Nr. 99, Potsdam 2005

Aviation Management

The Role of Business Aviation in the European Civil Aviation Market

Raimund Hosch

1 Determining the Position of Business Aviation in Europe

Business Aviation – being defined as the branch of general aviation that covers non-scheduled commercial, corporate or owner operated aviation services as an aid to the conduct of the respective business – gains more and more market share in the premium segment of the civil aviation market. What are the implications of this development for the established "scheduled" carriers? Are there any? And if yes – would these rather be opportunities or threats? Are there any business aviation specific benefits for the corporate traveller? Will the emergence of highly cost efficient microjets increase the economic viability of "cab flying" for everybody? Are these developments the first indicators of a fundamental paradigm shift in the civil aviation industry?

A panel of industry experts – Dr. Bernd Gans (CEO German Business Aviation Association), Dr. Hans-Joerg Hunziker (CEO JetBird), Mr. Peter Kaiblinger (Director Sales Germany of Embraer) and Volker Thomalla (Chief editor of FlugRevue) discussed these questions in a panel that took place during the ITB Berlin Convention in March 2007.

2 Trends in European Business Aviation

When talking about business aviation it is usually referred to aircraft from four to forty seats. Business aviation customers either require a very exclusive flight travel product and value personal discretion extremely high or they have very urgent time needs or a combination of both. For this customer group, rates and prizes are of comparatively inferior significance. Typical business aviation flights facilitate access to cities and regions that are beyond the reach of scheduled carriers.

The typical business aircraft has two engines and a maximum takeoff weight of 5.7 tons. Germany, for example, has a population of nearly 500 aircraft fulfilling these criteria. After all, in 2006 approximately eight percent of all European flights under IFR (Instrumental Flight Rules) were business aviation flights. Over the last decade the upgrading of business aviation aircraft has been one predominant trend. In 1996, two thirds of the aircraft taken into consideration featured turbo prop engines, only one third operated on jet engines. In 2006, this ratio has shifted to vice versa: Now two thirds of the aircraft in question have jet engines, only one third remaining with turbo prop gear.

Some major aircraft manufacturers strongly believe in the market potential of business aviation. One approach is to try to benefit from technological know-how in manufacturing regional or even bigger commercial jets and introduce "Light Jets" and "Very Light Jets" that are based on similar technologies. Embraer is certainly one of the most prominent aircraft manufacturers to follow that path, Bombardier is another. The former company started a product line of business jets five years ago and celebrated the hundredth delivery in March 2007. A new very light jet aircraft with four passenger seats and two engines has just been introduced to the market.

Actually, the newest generation of Very Light Jets features flight deck – particularly display and control-technologies that match the standards of state of the art double aisle aircraft. Besides, new jet engine technologies have dramatically reduced specific fuel consumption. Operating costs of current two jet engine aircraft have been coming down to those of turboprops. Routes that can be operated by jet aircraft in an economically viable way are decreasing. Whereas by the end of the 1990s the economically viable route for a two jet engine aircraft of the size in question was five hundred to seven hundred nautical miles, today this kind of aircraft can be operated profitably on short haul flights that a decade ago used to be the exclusive domain of turboprops only.

By flying point to point, the specific fuel consumption profile of business aviation tends to be relatively environment friendly: Flying direct by avoiding hubs and therefore detours equals more resource efficient flying. On the other hand it remains obvious that due to the essential law of economies of scale an aircraft with four to eight seats will always have fundamentally higher fuel consumption per seat than an A 340 or comparable aircraft.

Noise emission is another ecologically relevant dimension of civil aviation. Due to the relatively small size and weight of the fuselage and the performance of engines being used in business aviation aircraft, the typical noise pattern of business aviation aircraft is far smaller than the one of e. g. an A 320. One reason for this advantageous feature – "not bigger than a bed size rug" as one panellist puts it – is that a light jet requires a much shorter runway distance to reach take off speed than an aircraft the size of a B 737.

Only recently business aviation operators have been introducing innovative business models to customers. JetBird for example attacks the already established NetJets business model by a rate of 2,000 Euros per flight hour whereas Netjets levies a comparable fee of 5,000 to 5,500 Euros. Netjets business model is based on selling a certain quantity of user rights in advance whereas JetBirds model focuses on a "power by the hour" model. The operator's risk is therefore directly linked to the load factor that he is able to accomplish, leaving a comparatively low risk to the customer and therefore making this "Airtaxi" product very attractive. In order to be profitable this business model requires an annual minimum of eight hundred to one thousand flight hours per aircraft and a load factor of at least fifty percent. If these criteria are matched, a ticket prize, comparable to first class ticket prizes of scheduled airlines, becomes feasible. Netjets model is costlier for the customer, simultaneously reducing the operator's risk. Time will tell whether one or both of these models can survive in the market. JetBird obviously strongly believes in the Airtaxi model – the company has placed an order of one hundred Embraer Phenom 100.

Compared to scheduled airlines an Airtaxi operator is in a position to offer certain advantages to the customer. Airtaxis can operate flights between secondary airports – e.g. Munich – Lugano or St. Moritz – Le Bourget. Lead times at secondary airports are dramatically shorter than at primary airports: from curb to aircraft the average lead time is rarely exceeds fifteen minutes. Slots are easier available than at bigger airports. In many instances secondary airports are located closer to the residence of the customer than primary ones. Therefore significant time gains can be combined with more convenient direct flights. By using these secondary airports, business aviation airlines can benefit from a potential of eight hundred airports in Western Europe that are by nature too small for aircraft used by scheduled airlines. Actually ninety percent of civil aviation traffic takes place via only ten percent of the airports.

This topic still has another dimension, which can be illustrated using the example of Switzerland. Zurich airport, having a scarce number of available slots, is expensive when it comes to landing and handling fees. But a carrier like JetBird can tackle this situation by switching over to airports in Geneva or Basel.

The Western European business aviation market covers seventy percent of the whole market in Europe. The Western European business aviation network stretches basically from London in the north west to Rome in the south of Europe, therefore nicknamed "European Banana" according to its geographical shape. As Switzerland is being centrally located within this banana, it offers ideal conditions as a hub for business aviation airlines that want to serve Western Europe. Switzerland ranks number five as European business aviation country, with Germany being the number one market, followed by the United Kingdom, Italy and France.

In order to substantially improve load factors in business aviation, active sales efforts have to become far more intense than they are nowadays. New distribution channels and systems need to be established. There is a variety of possible strategies, which focus either on web based direct sales or on benefiting from extant travel agent/tour operator market structures, which so far sell predominantly conventional leisure or business travel products. For the weekend traffic –being generally driven by leisure travel motives- traditional tour operators seem to be the right partners. For the rather professionally motivated traffic from Monday till Friday, business travel operators like Carlson Wagonlit Travel or Hogg-Robinson could probably be appropriate sales partners for business aviation airlines. One important question on the technological side obviously is how business aviation products can be made bookable in the Computer Reservation Systems (CRS), which are being applied by all big agents and operators, to sell the vast majority of their travel products. In general, the implementation of comprehensive and consistent sales and marketing strategies still seems to be new ground for many business aviation operators.

3 Future Perspectives of the European Business Aviation Market

In the manufacturers view a demand for approximately 12,000 light and very light jets for the next decade seems highly probable. Looking retrospectively at the German market alone, an average growth of five percent per year took place over the past decade. There seems to be no fundamental doubt that this increase is set to continue in the future. On the other hand history shows that in the past the business aviation industry has seen downturns every seven to eight years and there is no reason to doubt that this cyclical character will disappear. Pilot training and licensing is also an important bottleneck. On the other hand it remains a fact that business aviation has enjoyed growth rates since 9/11 that were significantly higher than the ones in the scheduled airlines business. Due to globalization effects it seems more than plausible to assume that the need for business aviation products, particularly amongst members of the upper and middle management, will maintain its strong growth.

Airport Privatization and Takeover – Creating Value for All Stakeholders?

Adrian von Dörnberg

1 Airport Privatization Framework in the 21st Century

The past decade did not show a significant change in airport rankings. The top eight airports for passenger ranking in 2006 were Atlanta, ranking number one with almost 85 million pax, Chicago, London, Tokyo, Los Angeles, Dallas/Fort Worth, Paris and Frankfurt. However, four of these airports experienced a decline in passenger volume between 1.2 and 0.3 percent year-on-year, causing them to appear in the laggards list. In contrast, airports from emerging world economies are growing the fastest. Chinese airports continue to dominate the list of fast growers – and Delhi and Mumbai, India's main hubs, outpaced every other airport reporting a 20 percent increase in traffic in 2006. Obviously, one of the most important influencing factors for passenger traffic is the growth of the economy. In fact, airport privatization is all about growth and profitability. In the first section we will investigate what this means for existing business models and emerging new models for airports in the 21st century or at least over the next decade.

But how does Dublin (15 percent change) or Barcelona (11 percent change) fit in this picture of rapid double digit growth?

Ryanair, the fastest growing low cost airline in Europe, is "dominating" the Dublin airport with a 38 percent proportion of flights; Barcelona is experiencing tremendous traffic growth as a destination hot spot, especially among young people up to 35 years of age from countries throughout Europe. Traditional airline hub dominance based on frequencies can also impact traffic change and growth in an extremely positive or negative way.

Cincinnati had the misfortune of topping the laggards listing with a dramatic 28 percent decline in traffic last year (2006). Its fall is almost entirely due to Delta's decision in late 2005 to downsize its Cincinnati hub. It cut main line and regional capacity by 26 percent. Similar pullback in main line service occurred in Salt Lake City.

Hub dominance is significant for Houston (Continental frequency share of 76 percent), Chicago Midway (Southwest Airlines share of 72 percent) or London Stansted (Ryanair share of 59 percent), to name only a few. A pullback or shift in main line service to another airport results in a dramatic downturn or in a double digit growth for the airport which benefits from such a move.

Apart from pax growth, many airports experience cargo growth: In 2006, Asia-Pacific and Middle East airports reported double digit growth figures.

As growth is the backbone for airport privatization, it seems logical to study these figures: However, is volume the right indicator? Does such a growth in volume result in a proportional growth in revenues? And last but not least in profitability?

Is the environment, regulated and framed by governmental authorities, actually open for privatization at all?

To understand the complexity we have to have a look at the basics of the air transport system which consists of three major components:

- The airport as the physical site and point of interaction
- The airline
- The user

Each of the players must attain some form of equilibrium in respect to the others. Failure to do so leads to suboptimal conditions

- Infrastructure constraints
- Effects of environmental measures
- Exercise of traffic rights: "winners and losers"

All issues related to airport economics including private investments or privatization have to take the large number of stakeholders into account (Illustrated in Table 1).

Table 1: Organization affected by the operation of a large airport

Principal actor	Associated organizations
Airport operator	Local authorities and municipalities
	Central government
	Concessionnaires
	Suppliers
	Utilities
	Police
	Fire service
	Ambulance and medical services
	Air traffic control
	Meteorology
Airline	Fuel supplies
	Engineering
	Catering/duty free
	Sanitary services
	Other airlines and operators
Users	Visitors
	Meters and senders
Non-users	Airport neighbor organizations
	Local community groups
	Local chambers of commerce
	Anti-noise groups
	Environmental activists
	Neighborhood residents

Source: Ashford et al, Airport Operations p. 3

Los Angeles, London Heathrow and Atlanta, to mention only a few, have total site employment levels of more than 50,000. Some airports integrate shopping malls, cinemas, hotels, churches, auto-repair shops and service stations, hospitals, offices, spas, restaurants etc. An airport can be described as a medium sized town or suburb of a greater metropolitan area. The CEO of Amsterdam's Schiphol airport therefore calls himself 'the mayor' of Schiphol. First and foremost, an important player is the "non-user", who can have a very significant impact on airport operations and who is also greatly affected by large-scale operations.

Time (and therefore cost) consuming procedures are necessary to redesign an existing airport, build new runways and terminal capacities.

The new 'Capital Airport' in Berlin (Berlin-Brandenburg-International, BBI) provides a perfect case study for building a new airport and closing down existing sites (Tegel, Tempelhof). Villages had to be moved, the planning took more than 15 years, private investors/operators back out of the project. An ongoing political discussion by stakeholders is still influencing the project.

Fraport's intention to extend the runway system in line with a given company's expansion plan and Munich's similar plans for a third runway are good examples for the tremendous impact stakeholders have. These projects can today only be exercised or put forward within the framework of a regional planning policy and planning permission procedures with special emphasize on operative capability, environmental impact and economic viability.

The United States have by far the greatest number of airports in the world (Figure 1).

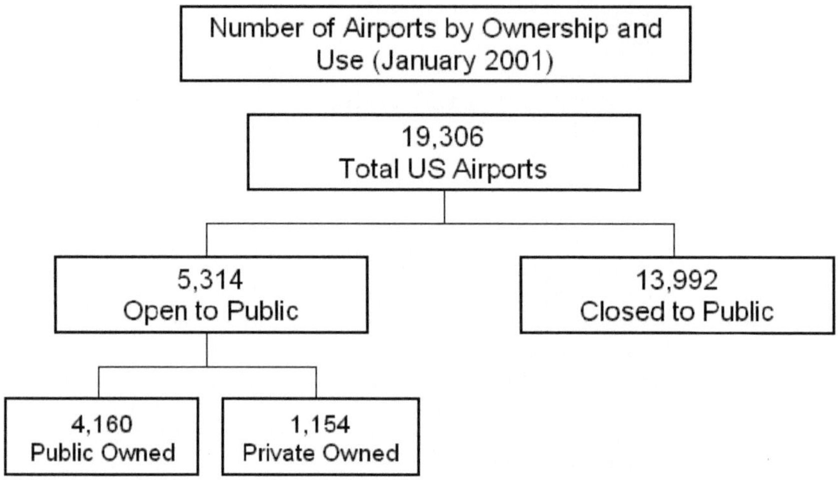

Fig. 1: Numbers of Airports by Ownership and Use; Source: Wells, A.T., Young, S.B.: Airport planning & management, p. 5

There are more than 19,000 civil landing areas in the United States, most of them privately owned, and for private use only. Another 5,400 airports are open for use to the general public; 4,200 are publicly owned, either by the local municipality, county, state officials. The remaining 1,200 are privately owned, either by individuals, corporations or private airport management companies.

2 New Business Models and the Paradigm Shift

Airline markets naturally evolve toward point to point flying over time. Since very few routes have enough demand to warrant direct flights, hub and spoke flights dominate the early stages.

As a market develops however, frequently travelled routes reach certain demand thresholds which justify direct daily services.

Although it may seem that the hub-and-spoke system will become less common over time, forecasts indicate that, in absolute terms, it will remain the predominant form of traffic over the foreseeable future.

To see what might drive this change, it is necessary to return to the discussion about aircraft technology. As airlines become more indifferent to the costs of flying a large or a small aircraft, a key plank in the logic for consolidating hubbing traffic into ultra-thick routes and mega hubs potentially falls by the wayside. In a bid to offer greater frequency and/or shorter flying times, airlines would in fact have a greater incentive to use smaller aircraft and disperse their hub flights across a number of routes and airports, rather than consolidating in a limited number of trunk routes flying out of mega hubs. Other factors may also contribute to loosening the ties between hubbing traffic and particular routes and airports:

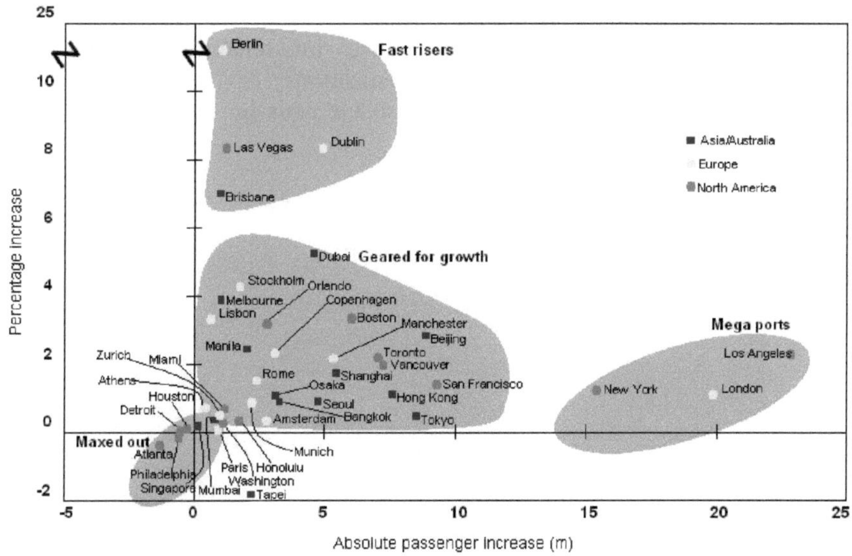

Fig. 2: Potential change in long-haul international passengers by city 2004-2015 *(Note: Growth based on estimated warranted long-haul passengers in 2015 compared with actual in 2004)*; Source: Boston Consulting Group

In face of increasingly open skies we will see carriers less inclined to deliver all their feed to one large alliance partner's home hub. Instead, airlines may bypass their alliance partners, and/or build codesharing relationships with low-cost carriers in other parts (secondary airports or designated low-cost airports). A "Boston Consulting" analysis shows a modeled comparison of "warranted" 2015 long-haul traffic of each major airport with its estimated long-haul volume today (Figure 2). Las Vegas, Berlin, Brisbane and Dublin for example, all stand out as cities that have outstanding potential for traffic growth in a splintered world. On the flip side some large hubs, such as Atlanta, Paris and Singapore, appear to have maxed out their growth potential. These projections will have an impact. A positive growth story is one of the driving elements for potential strategic investments in airport systems today.

Another positive driving factor is the fact that airports are highly profitable compared to other players in the aviation value chain – and in terms of profitability, privatized airports clearly out-perform those owned by the state; last but not least privatization also has an impact on operational efficiency (Figure 3 a-c). The traditional business model and the privileged position of airports are increasingly coming under pressure.

Airport revenue potential is threatened by

1. The airline industry (fragmentation, economic pressure, low-cost challenges, demands to transfer risk)
2. Competition (airport to airport: competing for connecting traffic, intermodal: competition for short-haul traffic)
3. Rising passenger reputations (convenience, time-reliability, customized offers, entertainment, information, communication)
4. Policy/regulation (environment i.e. noise and emissions, security, liberalization)

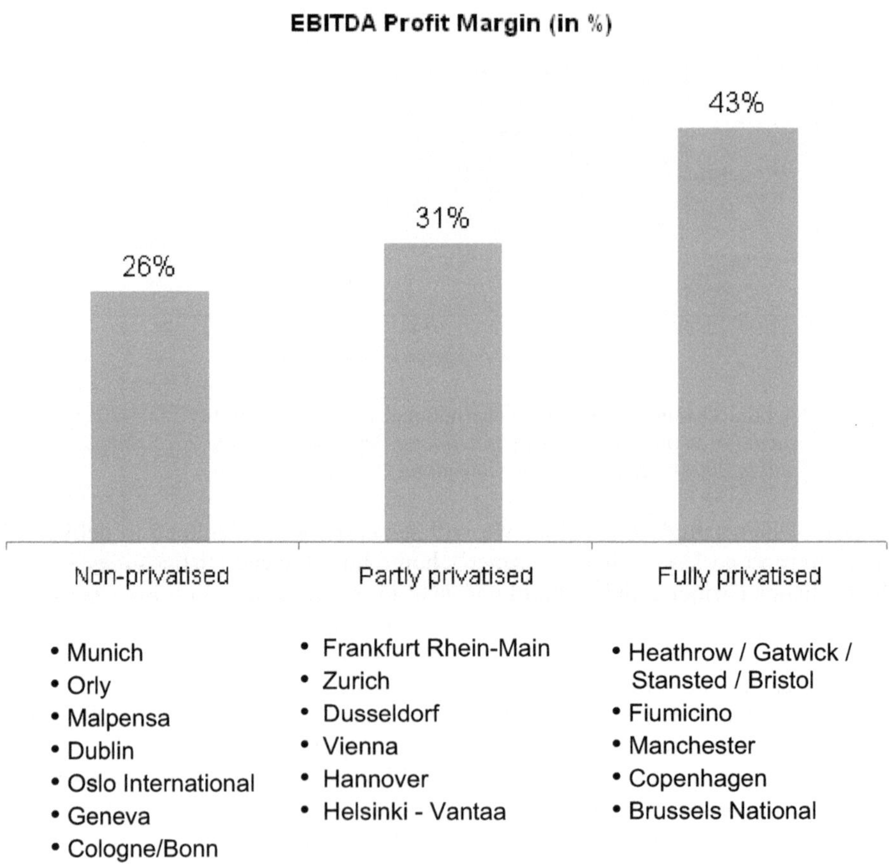

Fig. 3a: Source: Booz Allen Hamilton, presented on ITB 2007

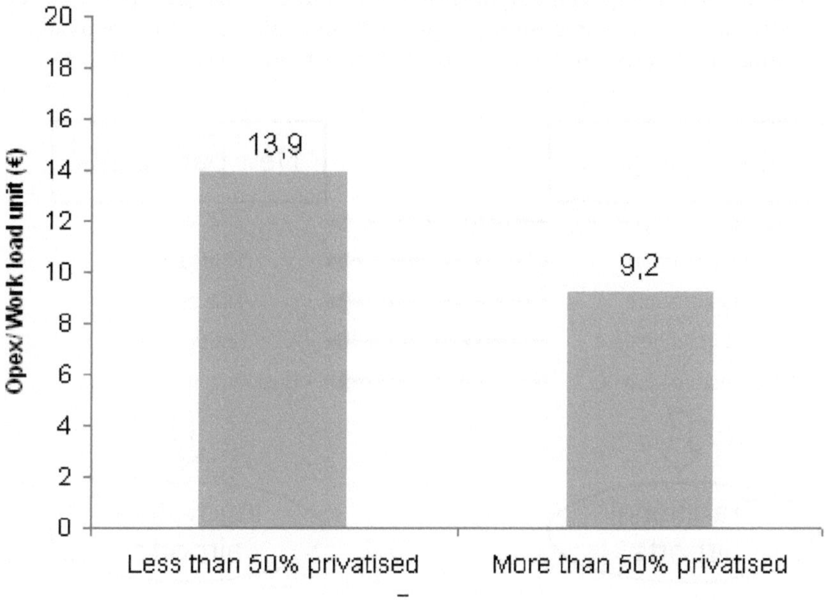

Fig. 3b: Source: Booz Allen Hamilton, presented on ITB 2007

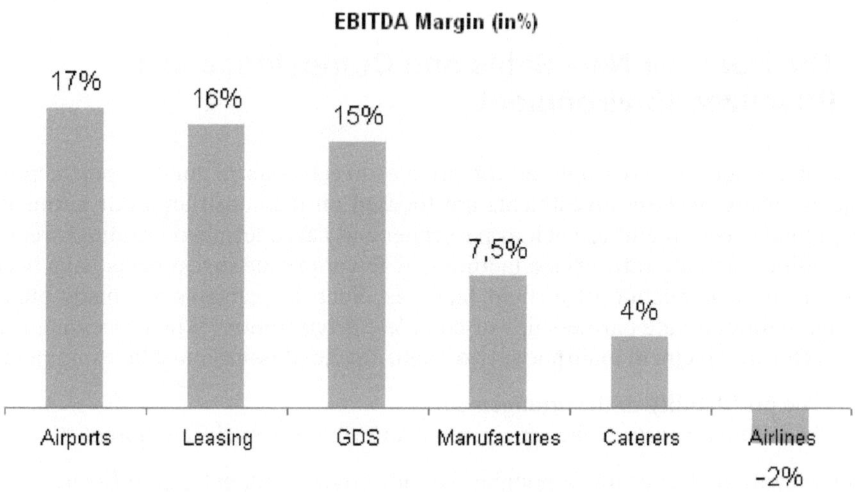

Fig. 3c: Source: Booz Allen Hamilton, presented on ITB 2007

Airports are undergoing a paradigm shift to face these challenges. The focus of the "innovative airport" lies in customers, profitability, value creation, innovation and – in contrast to architecture/layout – on a business model design (Figure 4).

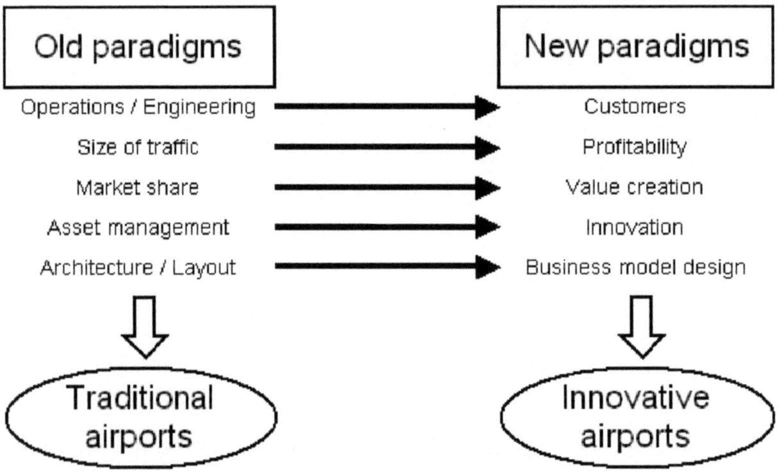

Fig. 4: Source: Booz Allen Hamilton, presented on ITB 2007

3 The Need for New Skills and Competence in a Privatized Environment

There is a wide range of options for private investments to fund airport capital projects. Many of these investments are focused on the construction of terminals and ground access facilities such as passenger and cargo terminal buildings, rental car facilities, and aircraft service facilities. A fewer private investments have been made in the construction of airfield facilities. Such investments are made either through public-private partnerships or complete privatization. When assessing private sector involvement in airports, two major financial issues must be examined:

1. The profitability of the arrangement
2. The ultimate costs of the arrangement and where the risks are borne

Privatization has become a popular way for government entities to finance new and existing infrastructure projects.

The private sector is bringing efficiencies to traditionally government-run projects. There are three generic forms of privatization today:

1. Build, operate and transfer contracts (BOT)
   Under a BOT-contract, private investments are used to construct and operate a facility for a period defined in the terms of the contract; at the end of the contract period, the ownership of the facility is transferred to the airport owner.
2. Lease, build and operate agreements (LBO)
   In a long-term lease (usually lasting from 20 to 40 years), the government allows the private sector company or consortium to build and manage an airport facility, while leasing the property and facility from the airport.
3. Full privatization
   The sale of the entire airport or partial interest in the airport is the final basic form of airport privatization. This form is prominent internationally. The government gives up all rights of ownership to the private entity; however, the government maintains it regulatory authority.

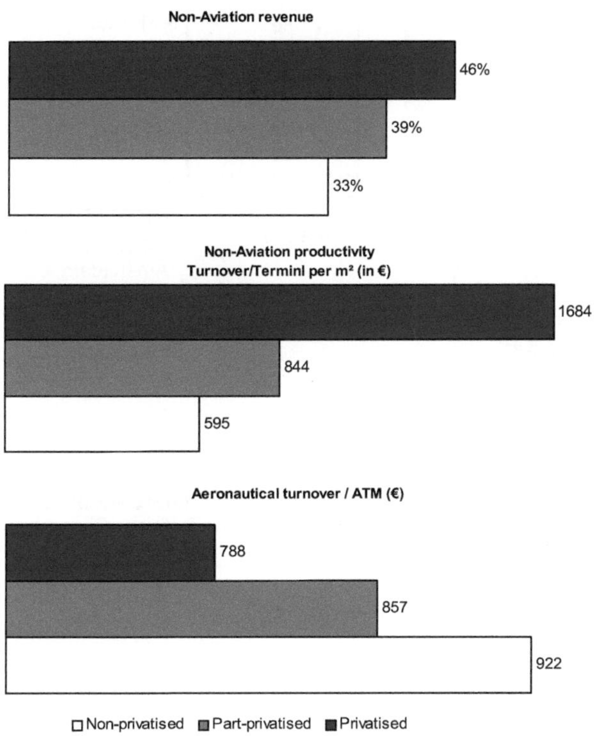

Fig. 5: Source: Booz Allen Hamilton, presented on ITB 2007

As shown in Figure 5, the differences in the business model also depend on the degree of privatization. Significant differences in "non-aviation-revenue", "non-aviation productivity" and "aeronautical turnover" underpin the new business model and refer to the term "innovative airport". An "innovative airport" can be described by the cleanly differentiated portfolio compared to the traditional ones (Figure 6). Innovative airports do indeed manage an often huge portfolio of business models: Shopping malls, business parks, entertainment centers, hotels, spas, cinemas and last but not least the core business which can be split in 'low-cost-terminal/airport activities', a cargo hub, a charter hub, an alliance anchor hub. In consequence of this development the managing director of Amsterdam airport calls himself "the mayor of Schiphol".

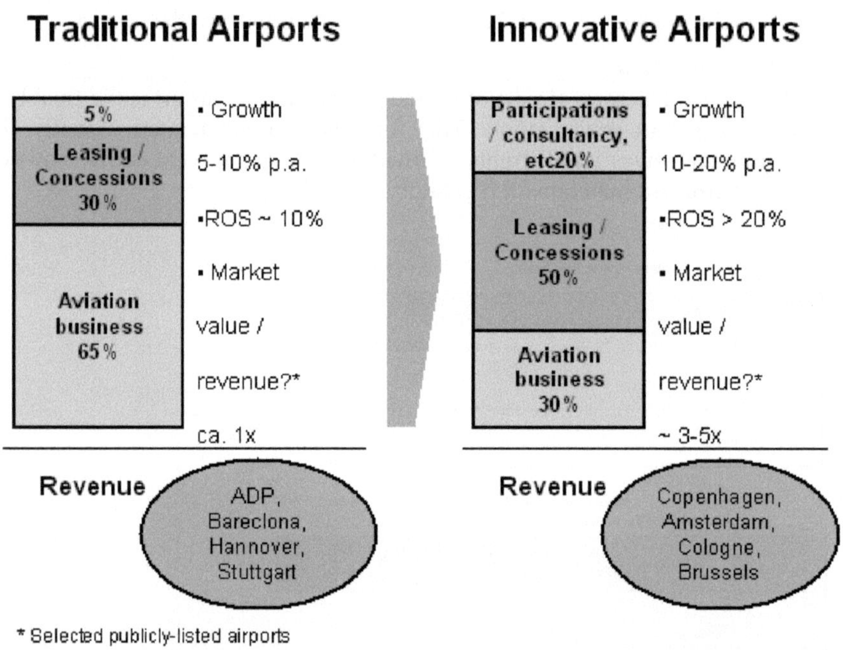

Fig. 6: Source: Booz Allen Hamilton, presented on ITB 2007

As analyzed before non-aviation revenue drives profitability. There seems to be no difference so far between an airport designated as "hub", "hybrid" (= combination of models) or "low cost" (Figure 7).

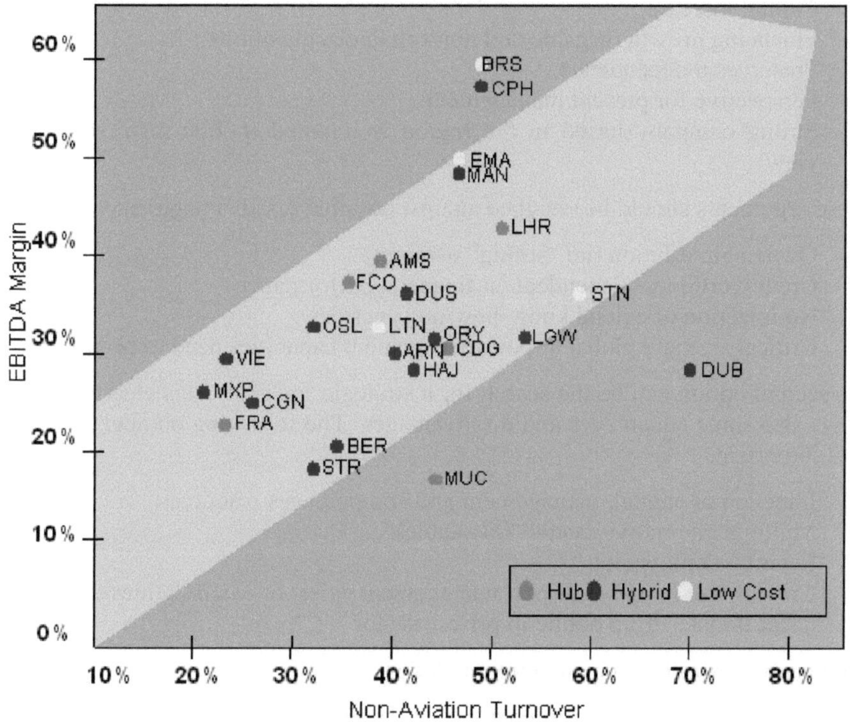

Fig. 7: Source: Booz Allen Hamilton, presented on ITB 2007

To manage the complex issues described above new management skills are required, especially to optimize non-aviation revenue. The claim should be to increase "time to spend". This means to increase share of time at the airport available for shopping. In order to do so, management has to reduce time at security to minimum, reduce boarding/time at gate to minimum and maximize "contact time" to service offerings. In a marketing sense or approach it can also mean to create incentives (for instance "vouchers") to stay longer at the airport and leverage knowledge to customer preferences.

4 The Main Issues on the Agenda for the Future of Airports

From the perspective of the airport run by government or state there are two generic alternatives to pave the way for privatization. The first option can be an IPO. For large airports with clear positioning and defined growth strategies, the advantages this option offers are quite compelling:

- Capital increase
- Financing growth (organic and non-organic acquisitions)
- Preserves independence
- Perspective for present management
- Strong company based in the region is retained (public resp. stakeholder view)

These arguments should be weighed against possible disadvantages such as

- Ownership dilution (no "strong" owners)
- Creditworthiness dependent on financial performance
- No injection of extend know-how/no synergies
- Critical size and global airport management capacities need to be built up

The second option can be the search for a strategic investor. This clear alternative shows also some advantages and disadvantages. The following number among the most important:

- Injection of capital, management and competences/resources
- Multiple alternative models (available)
- Synergy deployment
- Critical size in global airport management exists or can be achieved quickly
- Clear transfer from public to private sector

The disadvantages of such an option include:

- Capital strength of the company is not augmented
- Limited scope for existing management in planning for the future
- Large regional employer is no longer autonomous (public or stakeholder perspective)

Besides these challenges, classic marketing issues need to be addressed.

Figure 8 illustrates problems and approaches related to marketing issues: the so called AIDA-model can be used as a framework for the analysis.

Understanding the driving forces, consumer behavior and the global trends in marketing concerning airport management will be key to survive in the dynamic airline business. New players in Russia (Air Union as investor), the Middle East (Emirates) and China take the chance to position themselves as global leaders while governments of many developed countries are reluctant to open new windows of opportunities for growth.

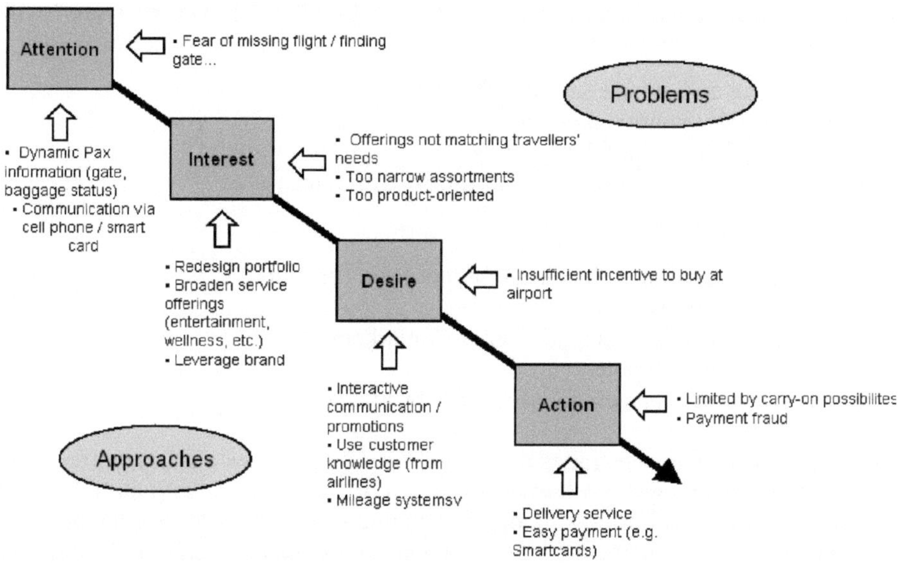

Fig. 8: Classic marketing problems need to be addressed; Source: Hauser, Richard, l.c

Understanding the driving forces, consumer behavior and the global trends in marketing concerning airport management will be key to survive in the dynamic airline business. New players in Russia (Air Union as investor), the Middle East (Emirates) and China take the chance to position themselves as global leaders while governments of many developed countries are reluctant to open new windows for opportunities of growth.

5 Conclusions

Airport privatization is still a hot issue on the agenda of the airline industry. New strategic investors are willing to pay very high prices.

Table 2

New strategic investors	Prices payed
- Ferrovial (BAA)	- BAA: 15 billion euro = 100 euro/pax
- Goldman Sachs/GE Capital (London City)	- London City:
Macquonmie Airports	1,2 billion euro = 600 euro/pax
(Brussels, Copenhagen)	- Copenhagen:
- Abertis (TBI)	2 billion euro = 100 euro/pax
- Hochtief (Dusseldorf, Hamburg, Athens)	- Budapest: 1,8 billion euro = 200 euro/pax

Source: Hauser, Richard : Airport privatization

Ideally, privatization creates a win-win situation for all stakeholders. "Users" will benefit from

1. Attractive and diverse service offerings
2. Significantly improved quality
3. Higher efficiency and lower costs
4. Injection of capital for necessary infrastructure measures

"Investors" will see

1. Protected position with "quasi-monopolistic character
2. High profitability
3. Low volatility and comparatively low market risk
4. Sustained growth potential
5. Paid customer contacts for secondary businesses

Regulation sets some constraints on profits, but also provides a safety buffer. Due to their revenue strength and low volatility airports continue to hold attraction for investors. The airport business model is changing significantly to face the increasing demand placed on them in a privatized environment. Increasingly airports need to develop new areas of expertise – alongside the traditional competences – above all in order to develop the non-aviation revenue. For large established airports with a clear strategic position, IPOs are an attractive option. Nowadays, as well as looking for capital, airports are principally also keep an eye open for strategic (re-)alignment and management competence.

References

Ashford N et al (1997) Airport Operation, 2nd edition, New York.

Hauser R Airport Privatization, Booz-Allen-Hamilton Paper and Presentation for ITB Aviation Day, Berlin 09 March 2007. (unpublished)

Borst S, Frank S, Kowalski U (2007) Abheben von überall. In: Focus. 30:122ff.

Leßner A. Wettlauf der Flughäfen. In: Travel One. 08.08.2007 p. 22ff.

Steinke S. Drehkreuz im Aufschwung (2007) In: Flug Revue, Februar 2007, p. 68ff.

Love, Ross / Goth, James: Decline of the megahub, in: Airline Business, October 2006, S. 72ff.

Häp, Ulrich: Flughafenausbau – Welche Start- und Landbahnvariationen kommen in Frage, in: Internationales Verkehrswesen, Nr. 9, 2007, S. 397ff.

O. Verf.: Daring to dream, in: Airline Business, June 2007, S. 40ff.

Ezard, Kerry: Heathrow reborn, in: Airline Business, June 2007, S. 48ff.

Rewe, Richard: Relief effort, in: Airline Business, June 2007, S. 52ff.

Pilling, Mark / Panarillo, Antonio: Table toppers – airports ranking analysis, in: Airline Business, June 2007, S. 57.

Wells, A.T. / Young, S.B.: Airport planning & management, 5th edition, New York, 2004.

Protocol (unpublished), Panel discussion about airport privatization; participants: R. Hauser, R. Schwarz, A. v. Dörnberg, Berlin ITB, 09. March 2007

Flughafen Wien AG, How to run an airport, Schwechat, 2007.

Hospitality Management

Hotel Architecture: Less Is More – Or: Natural Sustainable

Maria Pütz-Willems

Interview with Matteo Thun, Architect

Bozen, Florence, Milan… These three locations have shaped Matteo Thun's life. This architect and designer is now shaping the lives of many travelers. Studio Matteo Thun, located in Milan, has been providing innovative ideas on hotel ambience since 2004. Mövenpick, Radisson SAS, Falkensteiner, NH Hotels and several private hotels – they all commissioned this mid-fifty sensation to design their hotels which are now the talk of the town. And he has been very successful. The Side Hotel Hamburg was nominated for the prestigious "Hotel of the Year 2001" award, the Vigilius Mountain Resort near Meran in South Tirol, Italy, won the "Wallpaper Design Award" in 2004, and the Radisson SAS Frankfurt received a nomination at the Worldwide Hospitality Award in 2005 for "the best hotel opening of the year."

On principle, Matteo Thun challenges everything. It is in his nature to be a creative thinker. The South Tirol film director Traudi Messini exposed Matteo Thun in a very personal film: "You have to defend free thinking". Later Matteo Thun goes on to say that "[he is] always on the lookout for calculable risks." And: "Searching for durability is fundamental."

The Milan architect and designer is always coming up with surprises. At the ITB Hospitality Day 2007, tourism and hotel representatives discussed the changes in the Mediterranean and Alpine resort destinations. Matteo Thun threw a curve ball into the middle of the discussion with his statement "Design hotels are out and nature hotels are in!" Maria Pütz-Willems, the editor in chief at the online hotel magazine hopitalityInside.com and the moderator at the ITB Hospitality Day, conversed with Matteo Thun concerning trends and values in (hotel) architecture and design.

Pütz–Willems: Mr. Thun, your office creates designer watches, kettles, privately owned homes and hotels. The boundaries between architecture and design are getting blurry. Which of the segments do you prefer?

Thun: I always approach work in the same manner: I am an architect, who plans on large (architecture) and small (design) scales. It is a process, which I have learned as an architect: My approach is holistic – it takes everything into account from the floor plan to the acoustics and even the lighting design. Experts have labeled this approach "Milan School." Design should never be fury of form at any price.

Pütz-Willems: You have earned the reputation among investors to create profitable architecture. However: Do target budgets cramp your creativity?

Thun: I would never criticize an investor or their budget. I know the ropes of the game; that is all part of the job. I see your question nonetheless in relation to the structure of the resort hotels. The fundamental structure of a European hotel is intertwined with its "stand alone management". Most [hotels] are privately owned and family run – and are not affiliated with a global reservation system, but have all the advantages of a custom-made suit: They tailor vacations and even coordinate their regional location and guest needs. Large hotel chains hardly ever actually see a location's soul. The heterogeneity in the European vacation hotel industry is Europe's greatest asset. Hence, I fight for the "Jones" and "Smith", who are on the brink of a new generation, in which the youth choose to invest in new creativity, without abandoning their "stand alone" position.[1]

Pütz-Willems: Many resorts are choosing new modern designs because they want to attract international clientele. If people were to stay in these hotels, they would feel as if they were in a relatively comfortable business hotel. A fire may be crackling in the lobby, but instead of finding a traditional comfortable recliner with a foot stool, there is but an uncomfortable cube seat. Why do some designers not consider travelers' needs for rest and relaxation? And here I definitely refer to hotels in the luxury segment.

Thun: Hotel chains are subject to very complex price engineering. In fact, 80 percent of our job is price engineering. People such as I, could definitely make some changes in this respect. If a particular piece of designer furniture is too expensive for an investor, I can always design a less expensive alternative. There is an Anglo Saxon understanding of hotels and a Central European understanding. The latter propels the principle of individualization, not serial uniformity. Standardization is important for the price structure. Italians describe it as *La serie variata*. Within a series, they can change the clothes yet maintain the frame. For example: 100 police people are all wearing the same uniforms, but there are different "structures" underneath. Now the objective is to create identical bone structure so as to reduce clothing costs, yet design clothing which is much more individualized than ever before.

[1] Müllers and Meiers is analogue to the Jones and the Smiths.

Pütz-Willems: Have you lived in the hotels and rooms that you created? Do you test them?

Thun: Yes – whenever I am traveling, I do some "live" living and testing. On opening day, I am usually able to find hundreds of errors – an architect's eyes are critical and have to be much more critical than the consumers, the guests. However, my colleagues tend to "reassure" me because it is often only me who sees the small imperfections. For me though, it always remains a lesson for the next project.

Pütz-Willems: What architectural and designer lines are now "out" and what is "hip"?

Thun: To talk about "in & out" is nowadays mega-out. This is also true for hotel architecture. "In" is everything natural and homogenous, where the surroundings and building are in harmony inside and out. That is our goal.

Pütz-Willems: During the ITB Hospitality Day 2007, you basically exclaimed that "Designer hotels were out, natural hotels in!" How strong is the aspect of nature in you and your work?

Thun: In my statement at the ITB on designer hotels, I was referring to "superficially decorated hotels." Consumers, travelers, know a lot more about style and design now than they did a few years ago. Today, a hotel has to trigger and enhance wellbeing as well as be naturally aesthetically pleasing to the beholder. Terms change and wear out.

Pütz-Willems: What aspects does a Matteo Thun Hotel in the Alps or in the Mediterranean have?

Thun: Whether it is in the Alps or near the ocean – the architecture has to meet economic, aesthetic and socio-cultural demands from today and tomorrow. We always try to reflect a location's character and individuality in our projects: It all depends on the soul!

Pütz-Willems: Architecture and design embedded in nature – it sounds perfect. In practice, investors confront you with precise specifications for a property, the minimum number of required rooms, expected return on investment. How can a nature concept be calculated and integrated on such terms?

Thun: I am – as previously stated – accustomed to taking economic production processes into consideration and to holding a price-performance discussion with myself. A building, which interacts responsibly and intelligently with the available resources in regard to construction and maintenance, which adopts a completely natural harmony with its surroundings and in which people feel at home, has a high market value! A natural concept can most definitely be calculated: aesthetic and technical sustainability form a solid basis for return on investment.

Pütz-Willems: How have materials changed over the last few years? Does high-tech equipment go hand in hand with your requirements for naturalness?

Thun: Touch and warmth are important – we therefore use a lot of wood. It produces a desire for "high touch", the eyes' desire to touch. Natural and high-tech material can be successfully combined as long as the technology stays hidden. The art lies in treating the surfaces. We prefer to use wood from the surrounding areas; in the Vigilius Mountain Resort, we incorporated native larch wood. However, the essence does not originate in the choice of wood, rather in the technique, which is used to fall a tree. For example, "lunar wood" can only be cut in months which contain an "R".

We try to find investors, who are sensitive to such issues. It does not have so much to do with money as it does with conduct. Modern aesthetic is ethics.

Pütz-Willems: Could you identify some examples and figures for economically profitable natural hotel concepts?

Thun: I can give you a general description. Success does not lie in adding concept components; it is the art of subtracting. Less is more. Thus, using a clever floor plan, fewer square meters can yield higher hotel revenue.

This also signifies that the industry has to consider pre-fabrication. Little by little, prefabricated living spaces will come to be; they are quicker and more economical. However, it is important that it goes unnoticed by the customer. Such methods have been used in private housing for many years, and will also most likely have industrial applications. Once a certain volume is reached, we will probably also introduce pre-fabricated material on construction sites.

Pütz-Willems: How strong is the link between natural hotels and intelligent technology? Is this a contradiction?

Thun: No, absolutely not. Intelligent high-tech is high-tech with low-tech investments. For example: An electronic program activates five lightings scenarios when a guest enters the room. The investment is maybe 200 to 700 Euro for this. Initially though, many guests experience a few minutes of irritation. The first item on their list becomes learning, i.e. "TV watching" or "nightlight". They have to immerse themselves in the user manuals… That is exactly what I want to avoid! We keep the innovative switches, but insist on intuitive technology. High-tech has to harmonize with the processes. For instance, a small LED light hidden under the toilet makes it easier for sleepy guests to instinctively navigate to the restroom. Lamps like this cost about 10 to 15 Euro.

Pütz-Willems: From your perspective, how big is the target group for natural hotels? And what type are these guests?

Thun: The target group is enormous! In particular people from highly populated urban settings with hectic and demanding lifestyles, who are on the road a lot – they intuitively sense when the location, material and functionality are in harmony.

Pütz-Willems: When and where do sustainability factors come into play?

Thun: The term sustainability could almost be defined as the buzz word of the early 21st century. Sustainability is the talk of the town, not only in the economy or in politics, but also in the construction industry. Considering everything closely however, only one aspect of sustainability catches the industries attention, the ecological perspective. In this respect, the use of different resources throughout a building's complete lifecycle is balanced. This ecological balance corresponds to the materials used from production to demolition and even the essential resource needs for building management throughout the whole usage period.

The two other aspects of sustainability are often forgotten, the economic and the socio-cultural points. A building should maximize its potential to reduce maintenance costs, if possible even generate profits and should lose very little in value. On the other hand, it should also cater to the users' wellbeing in regard to health and comfort aspects as well as be aesthetically pleasing.

Buildings with true sustainability are only then given when optimal harmony can be reached between all three aspects.

On the job, we manage every project with this fundamental approach.

Pütz-Willems: Are investors willing to pay more for sustainable construction methods?

Thun: Buildings that withstand time and are built according to the aforementioned criteria do not need extraordinary budgets. We also supply the investor with energy balance for the first ten to twenty years.

Pütz-Willems: How much and which pleasures does your natural hotel concept offer? Which senses does it touch?

Thun: It touches all the senses. Our construction method ensures that material, proportion and space, lighting, and sound entice all the senses – people, the users are the protagonists.

Pütz-Willems: Does the natural hotel concept have what it takes to endure until 2020 or even extend beyond this period?

Thun: Precisely – yes!

Pütz-Willems: Should architecture determine wellbeing right from the outset, or even dominate it? To what extent can you address target groups and thus exclude others?

Thun: There will always be target groups, which do not 'fit' in this concept, which prefer the superficial – there are of course other concepts for such people i.e. in Dubai, Moscow, or Sharm-el-Sheik…

Pütz-Willems: In what type of hotel, or in which hotel model, can travelers find the highest amount of rest and relaxation?
Thun: True luxury is found in simplicity and the directness. We are currently planning and constructing many projects in the Alps, a landscape with which I am truly united. Direct contact to nature gives me the greatest amount of relaxation.

Pütz-Willems: How would you define a two-star Matteo Thun hotel, and a luxury hotel?

Thun: A two-star hotel has a nice bed, a nice shower – and possibly a coffee before departure. A six-star hotel has excellent service, a wonderful spa, a spectacular location.

Pütz-Willems: You are currently planning a budget hotel chain under the name of Zero Star? What do these hotels offer?

Thun: Zero Star is a concept, with which I have been toying for a long time. Many business travelers check-in at night usually after dinner – and are already on their way before the crack of dawn. Nothing is more important than a comfortable bed and a nice shower, in which the guest can still have room to move around. Functional, goal-oriented, yet nevertheless well proportioned and inviting: *Simply beautiful* is the Zero Star Motto. The name alone should already suggest that categorizing such a hotel is unnecessary. We focus on simplicity and smallness, on *small is beautiful*. In practice, this concept is also defined as: One designer chair in a room is simply not enough. That is bluff. I continually struggle against such preconceived notions.

Pütz-Willems: Where will Zero Star Hotels be located in the future?

Thun: In "triple B" locations, in Europe and abroad. We already have five projects in the planning phase. Specific plans will come to light in the spring of 2009.

Pütz-Willems: How does Matteo Thun travel? Fast, slowly, comfortably, goal-oriented? In which hotels or hotel types to you prefer to stay?

Thun: Unfortunately, I am in the sky a lot – even though I do find it quite pleasant to take the train, where I can read, use the telephone, and completely abandon myself to ideas and their implementation. I find peace in old wise walls … i.e. the Ritz in Paris with a view of La Place Vendôme or the Sherry Netherland in New York with a view of Central Park.

Thank you very much for the talk, Mr. Thun!

Financial Solvency and Credit Rating for Hotel Businesses

Werner Pauen

1 Introduction

Business financing is still undergoing a transformation process: In recent years, increases in non-payment of credit and reorganization in equity deposits in credit institution as a result of Basel II led to a more restrictive capital market (Paul et al. 2002). Current developments on the capital market point toward a "credit crunch" in various economic sectors. This change is especially relevant to companies in the tourism industry:[1]

- For the most part, these companies are small and medium-sized businesses (SMB). Although many stipulations were made in favor of SMB during the Basel II negotiation period, it is to be expected that financial institutions continue to focus on the "good" clients and overlook the "bad" clients, such as startup entrepreneurs and medium-sized businesses, the SMB. A principal reason for this treatment is the small equity base held by SMB in comparison to the large corporations (Klein, Anclam 2002).
- A good portion of the main financing source for SMB is borrowed capital, or more precisely bank credits (ibid.).
- Available credit estimates for tourism businesses are often unfavorable.[2]
- The reservation of creditors which exists toward tourist infrastructure projects complicates credit approval even further: Reasons for this reserve are the specific risk factors attached to tourism (Bernet, Bieger 1999: p. 12). These aspects include the substantial dependency on seasonal fluctuations and changes in consumer behavior.

---

[1] According to the definition of the World Tourism Organization , tourism includes all activities related to persons traveling to a location away from their usual surroundings for leisure, business or other purposes for a period no longer than one uninterrupted year. The WTO's definition includes personal and business trips. See WTO 1998, p. 13-14.

[2] See the section *Credit Assessment Approaches* in this article.

The methods and procedures used in credit rating to measure financial solvency are still very important for businesses in the tourism industry. The less the specific character of this industry is considered, the greater is the risk of negatively assessing the credit situation of an individual company despite its actual good standing for profit and growth.

Credits still dominate business financing and are thus the resulting modus operandi for credit ratings for credit beneficiaries.[3] A variety of approaches are implemented, from simple account analysis to mathematical and statistical methods or even rating systems. They strive to define the credit situation of a business while taking quantitative (based on past performance) and qualitative (based on future estimates) factors into account. These analyses, which occur as partial analyses or integrated processes, should take into account as many credit factors as possible. In practice, this is where the rating process reaches its limits. The following paragraphs will deal with such analyses and an alternative scientific approach for rating credit factors for tourism and hotel businesses (Pauen 2007a).

2 Credit Assessment Approaches: Ratings

Credit analyses count among the relatively weakly structured areas of research. The decision-making process is so complex that defining the decision-making problem itself or determining a universal problem-solving method is next to impossible (Eigermann 2001: p. 51). Thus, the theory shortfall also encompasses identification of credit relevant information. A "theory on credit analysis" describing a cause and effect relationship for corporate characteristics does not exist.

In practice, credit assessments are usually carried out using a rating system. A comparison of prevailing rating approaches within German-speaking countries indicates that rating decisions are made based on the assessment of quantitative and qualitative factors. In addition, they usually introduce a supplementary assessment designed for a specific industry.[4] Assessment processes, within the scope of traditional quantitative analyses, focus on financial statement analyses. This alone does not do justice to hotel businesses. Their credit depends heavily on qualitative factors: They shape the monetary image of a business, including the financial statement, the profit and loss statement as well as the cash flow figures. In comparison however, none of the different rating processes indicate a uniform checklist which reflects these qualitative data.

The rating based on audited year-end figures can be performed using objective credit assessment processes, such as discriminant analysis[5]. Conversely, choosing

[3] In Germany, the equity ratio for businesses is 20-25% on the average. In comparison, Anglo-American businesses report ratios of approx. 50% (Kleiner 2001: p. 11).

[4] A detailed comparison of rating processes in German credit institutions can be found in Eigermann 2001: p. 120-120.

[5] "An overall rating process is only objective if the choice, weight and context of the rating criteria is independent of the subjective perception and experiences of the financial analy-

and assessing the qualitative factors, such as management quality and customer relations, are highly subjective[6]. They depend on the experience, specialized knowledge, power of deduction, and intuition of the decision makers (Eigermann 2001: p. 87). This subjectivity intensifies in correlation with the increasing complexity of the intertwining factors and mounting array of information. The rating process can explicitly define an industry-wide assessment which encompasses specific (future operational) factors of a business. Aside from comparing the results of the quantitative analysis with other companies within the industry, an analysis of the industry can also be performed. This includes the evaluation of market growth, concentration, market entry conditions, the degree of competition, and secure resources (Keiner 2001; p. 317). This information can either be incorporated into the global rating or included as a component of a qualitative partial rating.

However, an industry rating per definition does not really possess the capacity to assimilate company specific characteristics within the credit assessment.

The explanation: The industry rating merely indicates to what extent affiliation with a specific industry contributes to the general business risks of an individual company (Keiner 2001: p. 317ff.). This process only helps to ensure that a business assessment proceeds analogue to the competitive environment. When performing an industry rating, it is impossible to forego a specific risk analysis for an individual company based on the credit criteria (Büschgen and Everling 1996: p. 335).

This fact becomes evident when examining popular industry assessments. On a scale from "AAA" (the best ranking) and "E" (the worst ranking), in the period from 1999 to 2004, the hotel sector (excluding hotels serving only breakfast) received ratings between "C" (2000/2001) and "D" (2004) with a global rating stating that it was "significantly above the industry's average risk level" (2004) (Feri Branchenrating Deutschland 2004).

If such a negative industry evaluation were to be incorporated into a company's credit assessment without considering the specific qualities of the company, the resulting overall rating would not accurately reflect the company's situation. Qualifying a whole sector as a "sector in crisis" can result in negative ratings for individual companies, even though they have an excellent market position (Kley 2003: p. 178). Therefore, the question arises as to how the hotel businesses with good credit rating can be distinguished from those with poor credit rating.

sist on a widely empirical basis and if the methods followed are of mathematical and statistical value." (Baetge et al. 2004: p. 496)

[6] In this case, qualitative factors denote economic elements, which cannot simply be taken from the financial statement (Eigermann 2001: p. 64).

3 The Quintessence for Credit Ratings in Hotel Businesses

A rising tendency in credit criteria research, in respect to rating criteria, is to investigate ways to integrate qualitative factors into the credit rating process using objective perspectives. On the other hand, in practical applications, credit reviews are still predominantly based on analyses that use year-end figures (Eigermann 2001: p. 26). There are various explanations for this divergence:

- It is still difficult to identify qualitative business specific credit criteria.
- In order to process qualitative credit criteria with a standardized process, they first of all have to be quantified – a complex and time-consuming procedure.
- And ultimately, a qualitative database is often inapt for use with methodic statistical analyses.

In addition, from the financial perspective, detailed, specific industry know-how is essential. Therefore, a creditor working in the tourism and hotel industry should have the capacity to assess market and product concepts, positioning strategies and investments (Bernet, Bieger 1999: p. 47).

In respect to these circumstances, a credit analysis, which incorporates specific factors from the hotel industry should meet certain specifications:

- *Specification No. 1*: The choice of qualitative factors specific for hotels has to be logical. This requires a suitable model.
- *Specification No. 2*: The significance test for the selected factors has to be carried out using statistical procedure. Qualitative information should be logically associated with quantitative information and summarized as credit evidence (Keiner 2001, p. 142).
- *Specification No. 3*: Developing a databank specifically for hotels is also essential. This information would include figures from financial statements as well as the hotels' (qualitative) characteristics with relevant credit ratings in quantified form.

3.1 Specification 1: The Business Model in the Hotel Industry as an Analysis Unit for Determining Credit Relevant Aspects

Due to the different strategic orientations in the hotel industry and the specific quality standards and the degrees of standardization, it is very important to differentiate different hotels as much as possible. The array of hotels is far reaching: from individual luxury hotels (relatively high standards of quality, relatively low standardization for service) to one-star hotel chains (relatively low standards for quality, relatively high standardization for service), see Figure 1.

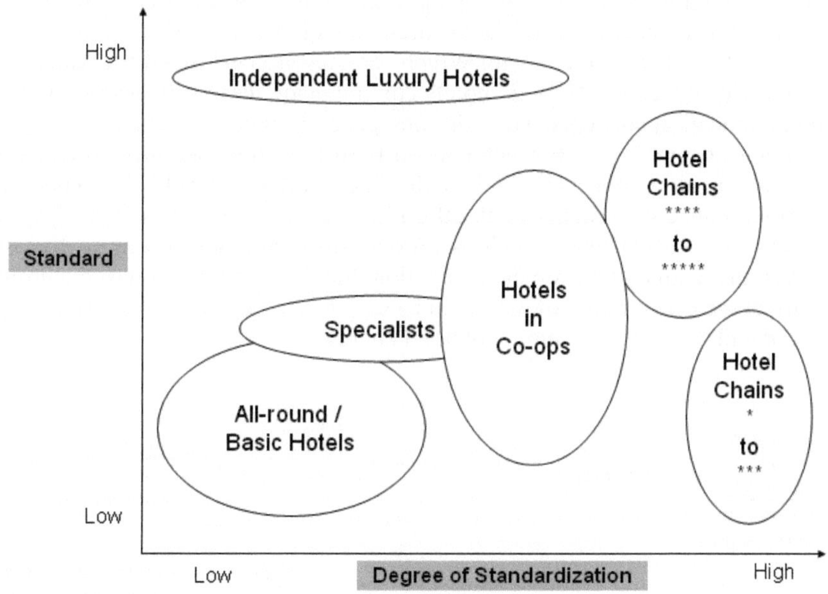

Fig. 1: Business models in the hotel industry (Source: TREUGAST 2004)

Measured by the overall bed capacity, hotel chains and hotel co-ops maintained 325,000 beds in 2005, which accounts for 36.4% of the overall room capacity in the German market[7]. They are thus to be considered principal providers and are also one of the strongest growing organizations in the hotel sector[8]. Hence, the following presentation will be oriented towards hotel chains.

How can the specific credit relevant factors in hotel chains be logically identified (Specification 1) in order to test their credit relevance? This paper makes use of the business model approach to respond to this query.

Departing from the operational synthesis in the theoretical approach for the "business model principle", concurrent structural elements can be determined. Based on this information, business models can then be developed.[9] These components supposedly facilitate "the presentation of the way in which a company, a business system or an industry generate value on the market".[10]

---

[7] IHA Hotelverband Deutschland, Hotelmarkt Deutschland 2006: p. 68. The Internationale Hotelverband Deutschland (IHA) summarizes hotel chains and hotel coops under the concept of "branded hotels". The basis of this definition is to present guests with a standardized hotel brand (ibid.: p. 60). However, in regard to credit assessments, the author believes that it is important to distinguish between these sub-sectors of the hotel industry due to their different business models.

[8] Compare with Pauen 2007b.

[9] Compare with Bieger et al. 2002: p. 46f.

[10] Compare with Bieger et al. 2002: p. 50.

This definition is used in the following discussion as an analysis instrument for hotel chains for the net effects of the business model. As such, the responses to the ensuing questions take center stage: Which services are performed for which customers (service concept)? How do hotels communicate these services on the market (communication concept)? How do they generate revenue (revenue concept)? How do they ensure growth (growth concept)? How do they organize core competences (organization concept)? And, with which partners should they cooperate (cooperation concept)? (Bieger et al. 2002) In a second stage, the extent to which the factors of each concept – such as product differentiation in one to five star categories, the degree of customer satisfaction, the financing method or the growth rate – affect the credit rating of a hotel in a chain has to be determined. To sum up, Figure 2 details the business model in hotel chains.

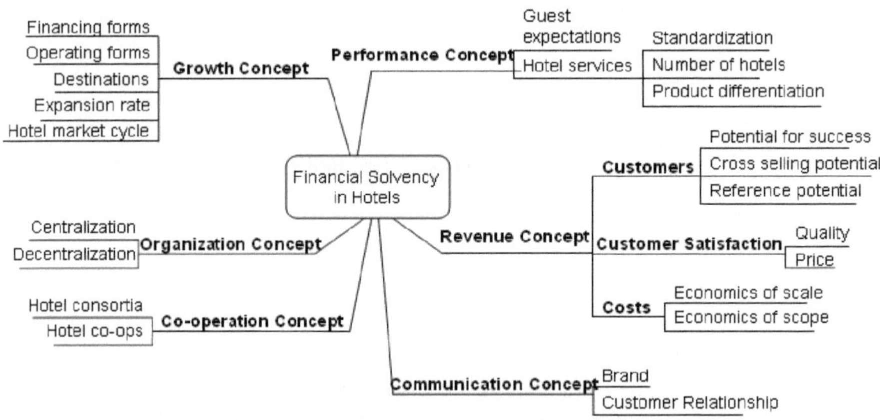

Fig. 2: The business model for German hotel chains (Pauen 2007a: p. 56)

Thus, the business model becomes an instrument for analysis. This model identifies the qualitative factors, which influence a business's credit rating.[11] Nonetheless, it is important to keep in mind that the credit relevant issues within the business model can vary over time.

Discussing all issues with credit relevance is not possible within this article. Some considerations are however presented for hotel chains. For example, customer satisfaction within the revenue concept for the hotel chains provides relevant areas of research in terms of financial solvency and credit rating.

[11] To view a detailed version of the hotel chain business model consult Pauen W (2007a).

Customers are key players for value enhancement potential in the service industry. They

- continually return (success potential),
- access additional services (cross selling potential),
- communicate (positive) experiences (reference potential), and
- contribute to product and quality improvement with feedback (information potential.[12]

Therefore, it is the sum of all customer values which influence the success of a business in the service industry. Customer values are dependant on customer satisfaction. In keeping with the classic target-oriented business objectives, these customer satisfaction are in fact to be regarded as key subject for managing and benchmarking customer-oriented business management (Kaiser 2002, p. 186; Hinterhuber 1999, p. 3ff.). In this sense, customer satisfaction management begins by knowing that customer satisfaction represents basic business survival (Matzler and Pechlaner 2001).

Although customer satisfaction for (monetary) success in a service business is crucial, there have been almost no scientific methods within the hotel industry, that have investigated this correlation, until just recently. Such empirical studies exist only in related industries such as the airline industry.[13] In the hotel industry in particular it seems that only using strategies like cost leadership increasing profitability has very slight chances of success.[14] However, this increased profitability can only be attained if the quality-price ratio corresponds to customer expectations.

The correlation between financial solvency and customer satisfaction can be seen in the investigation from Matzler and Stahl (Matzler and Stahl 2002). According to these authors, customer satisfaction and key financial figure "cash flow" are in a causal relationship (see Fig. 3).

[12] Consult Diller (2002), p. 162.

[13] In his investigations on the airline industry, Wricke identified a positive correlation between customer satisfaction for the product and the customers' willingness to pay (Wricke 2000: p. 162).

[14] Compare with Adam et al. (2002) pp. 762–763; Diller (1997) p. 749ff.

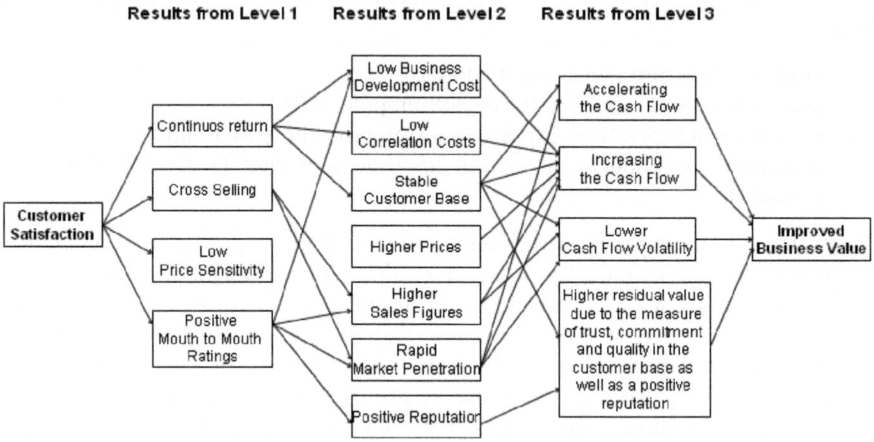

Fig. 3: Customer satisfaction and business value (Matzler and Stahl 2002)

Another example for possible questions relevant to financial solvency in the hotel industry are investments in brand names, which counts as one of the central issues in a communication concept for hotel organizations. "Branding" in a hotel chain is quite important, fulfilling the role of an information medium about the material and immaterial characteristics of a hotel product. From a hotel group perspective, the objective of a brand name is to create a standard within the hotel chain that is to be upheld in all hotels, providing consistent quality for service and design, and thus creating an overall corporate image for the customer (Frei 1999: p. 104).

Hotels in hotel chains use the brand as a stepping-stone to differentiate themselves from the competition. To achieve this distinctiveness, hotel groups have to maintain a sustainable brand image that reflects the characteristics of the services offered within the brand.

Expenses and investments in the brand are directly related to liquidity during the phase of brand development. Their benefits are ambiguous at first and are only perceptible in later phases. In addition, the expenditures for brand development can only be entered as an asset in certain cases – depending on the chosen accounting standard.

Furthermore, can the effects growth strategies of a hotel group have on the company's financial solvency also be investigated? The German hotel market is very dynamic on the supply side. The last several years were characterized by rapid growth in the number of hotels. Thus, it seems evident that one should also evaluate the driver of growth in regard to its financial solvency properties.

The largest growth potential for the hotel (chain) industry arises from the expansion of number of hotels. Which feature of the former and planned expansion is currently influencing the credit ratings? Does a diversified product strategy with one to five star hotels for example promise better credit rating in the long term than concentrating on the high-end segment in key business trip destinations? If so, to what extent or with which ratio do the individual segments affect the overall hotel portfolio? In reference to a balanced rating, can growth be easier attained in owned hotels or in management hotels? How does the speed in which a hotel chain rolls out new hotels affect the credit rating? In this regard, should a fast-growing hotel corporation be positively rated on principle? How much do external factors – economic trends – affect the credit rating?

Traditional credit analyses do not provide answers to any of the questions presented in the business model – as for example in the previously discussed aspects of customer satisfaction, branding and company growth.

3.2 Specification 2 and 3: The Significance Test with a Specific Database for Hotels

Business activities within each of the valid business models impact the financial solvency of the hotel business with their specific characteristics – such as financial forms, the expansion rate, the degree of product differentiation, customer satisfaction, and brand development. Financial solvency can be measured with rating indicators. Basically, in the selected approach, these indicators are composed of figures from financial statements, which consider proceeds, cash-flow, equity and outside capital as well as the total assets (see Figure 4).

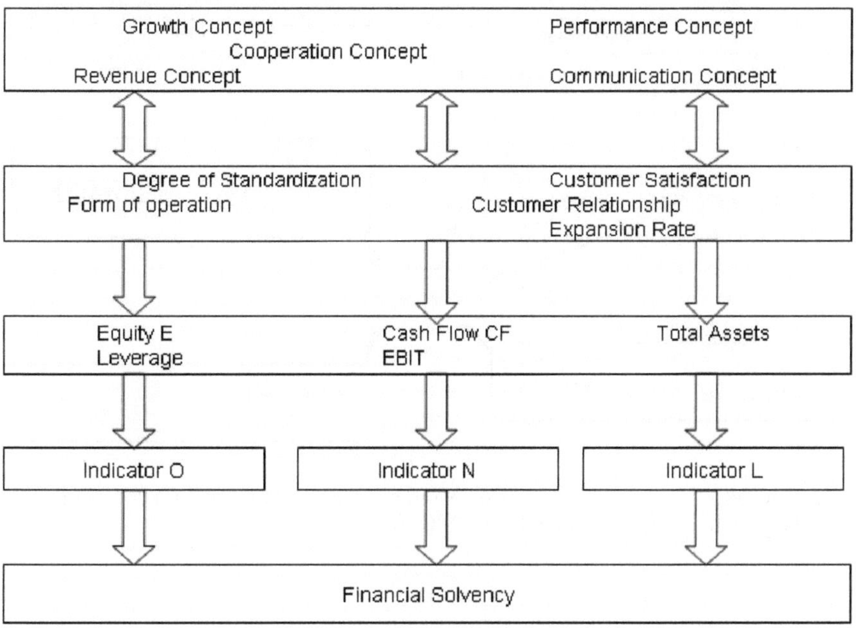

Fig. 4: Credit Rating Determinants (Source: designed by the author)

One of the analysis processes applied to these correlations is for example presented using economic effects and internal growth drivers (see Fig. 5):

- The influencing factors based on the business model, i.e. the expansion rate (number of hotels per year) or the economic cycle for the hotel market (number of overnight stays (Fig. 6)), create a functional correlation with quantitative credit indicators.
- Only rating processes, which are deemed objective, are used to identify financial solvency indicators. As such, the most developed processes are used – i.e. discriminant analysis, logarithmic regression and a neural net process.
- The financial solvency indicators presented in the model are based on financial statements from actual businesses in the hotel industry.
- Individual factors in the business model are consulted *ceterius paribus* as descriptive variables. At this juncture, economic development introduces the parameter "number of overnight stays in the German hotel market". The internal growth criteria are quantified using the "number of hotels" and the financial effects of "leasing and interest payments for hotel expansion".

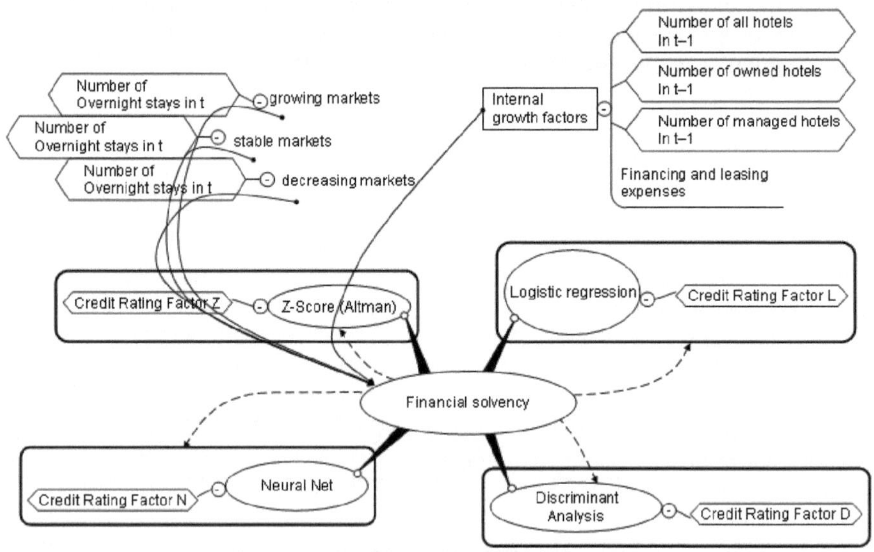

Fig. 5: A model of a credit rating analysis based on a specific business model – in this case the growth concept (Source: Pauen 2007a)

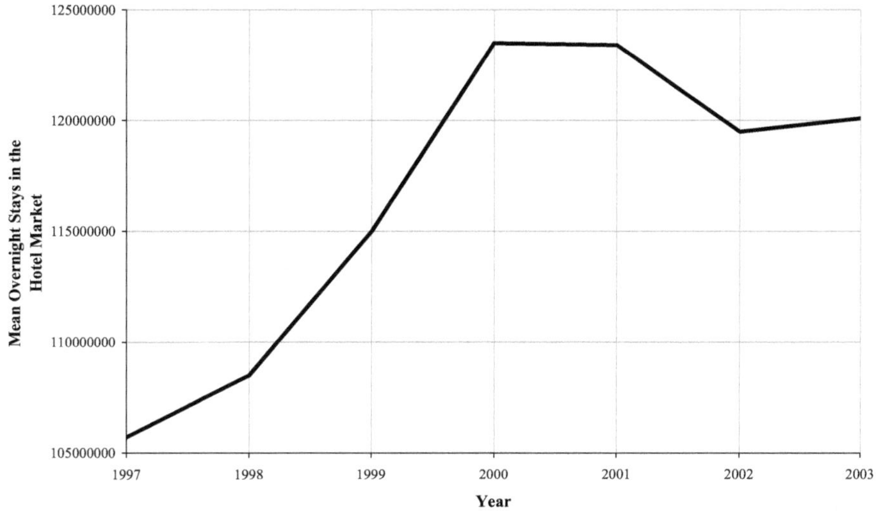

Fig. 6: The development of overnight stays in the German hotel market (Source: Statistisches Bundesamt, 2004)

In order to carry out the analysis, a database specifically for hotel chains had to be developed. All of the quantitative and qualitative values required for the analysis were loaded into this database.

In addition, a suitable statistical process had to be selected, which could also interpret small amount of data. The financial statement data, which are the fundamental financial solvency indicators, are only issued once a year on the balance date. Hence, there is never a sufficiently long time series to perform calculations. The panel analysis, a regression evaluation process, was chosen since it also provides valid data with short series. Businesses in German hotel chains are represented in the panels profile. To improve the significance of the analysis results, the panel analysis is performed on the same database for different financial solvency indicators *D*, *L*, and *N*.[15] Equation (1) expresses the correlation between the individual panel analyses.[16]

---

[15] As a compromise, emphasis was set on using a database, which contained as many financial solvency indicators as possible (over time) instead of a large range of businesses. See Hamerle A, Rösch D (2003) p. 206.

[16] The financial solvency indicator *D* and the exogenous variables, used in the six panel analyses, are set as an example. The equation is similar for the indicators *L*, *N*, and *Z* and all of the other independent variables.

$$D_{it} = \alpha + \beta_1 * HF1\_L_{it} + \beta_2 * HG1\_L_{it} + \beta_3 * HF1HG1QQ_{it} + \beta_4 * K_t + v_i + \varepsilon_{it} \qquad (1)$$

D_{it} = Financial solvency indicator D for the business i at time t.

$HF1\_L_{it}$ = A logarithm of the number of hotels (leasing and privately owned hotels) for the previous period t-1 for the business at time t, in other terms $HF1\_L_{it} = \log(HF_{i,t-1})$.

$HG1\_L_{it}$ = A logarithm of the number of hotel for the previous period t-1 for the business at time t, in other terms $HG1\_L_{it} = \log(HG_{i,t-1})$.

$HF1HG1QQ_{it}$ = The ratio for owned hotels to the overall number of hotels (leasing, privately owned, management and franchise hotels in the previous period) for the business i at time t, in other terms HF1HG1QQit = $((HF_{i,\,t-1}) / (HG_{i,\,t-1}))$.

K_t = The number of overnight stays in t.

v_i represents the business related influences.

ε_{it} corresponds to the time dependent and other uncontrolled influences.

4 Conclusions

The results, which were attained from evaluating growth factors in hotel chains through panel analyses, present the following picture:

- There is a highly significant negative correlation between financial solvency indicators and the number of hotels from the previous period and a positive influence on the number of overnight stays in all of the financial solvency indicators. In contrast, there was no evidence that the ratio between the number of owned hotels to the overall number of hotels had any affect on the credit rating (Panel 1).
- In an additional panel analysis, the effect the initial financial solvency in hotel businesses in 1997 had on the financial solvency indicators was evaluated. At this juncture, there seemed to be no consistent cause-effect correlation between the financial solvency indicators (Panel 2)
- In contrast to the earlier analyses, instead of applying the absolute number of hotels in the next phase, the financial impact of these hotel businesses was used in form of leasing and interest payments (Panel 3). The results attained from Panel 1 were both confirmed in respect to the economic indicator, *the number of overnight stays*, and the leasing and financing expenditures.

Using the same method, the other business model components in the hotel industry (revenue concept, service concept, etc.) can be tested in regard to their effects on the financial solvency. In order to quantify the corresponding business model components, expense and profit components for brand development or customer satisfaction can for example be consulted. Furthermore, financial solvency reviews can be developed for other sectors in the hotel and tourism industries or even for specific destinations.

The goal of this article was to reveal that the factors influencing financial solvency and credit rating in the hotel industry require specific assessment. In fact, combining the business model approach and the statistical method is a new and innovative method for determining and verifying credit rating. In this context, seeking an alternative approach does not necessarily have to be viewed as an affront to the rating approach used in the financial environment: Outside creditors in particular have to first of all make a compromise between their need for information and the costs required to generate and present this information when granting loans. Secondly, they must also provide employees with a practical assessment approach for day-to-day business dealings. In this respect, the approach presented by the author is to be viewed as a supplement to current assessment practices.

The approach based on the business model credit rating analysis – which is presented in an abridged version – should be continually revised and improved in regard to:

- the business model, which is subject to continual changes within its environment and thus needs to be redefined periodically,
- the factors of financial solvency, which are to be incorporated into the review process – these continual adjustments affect not only the selected processes (dicriminant and regression functions) but also their composition,
- additional developments in the business model database.

Nonetheless, it is apparent that the complex structure for financial solvency in tourism and hotel businesses has to be more transparent. This knowledge also holds true for direct loans for hotel businesses and for mortgages on the operating company's creditworthiness. Based on conventional analysis methods, the reluctance observed on the part of creditors and investors to finance tourism businesses and real estate is on one hand understandable in view of the fact that the hotel and catering industry has a relatively high proportion of insolvency, comparing to other industries. On the other hand, many well-positioned businesses, operating in the hotel market, promise investors and banking institutions an interesting rate of return or reasonable interest or amortization payments.

Therefore, taking additional information about the cause and effect relationship into account for hotel financial solvency is inevitable. This assessment should break away from a single-sided analysis based on financial statements and incorporate factors that influence a business's financial figures.

The innovative approach detailed here offers possible solutions. It reveals several factors relevant to a business's financial solvency which goes beyond financial, expense and revenue statements. As such, the financial situation of a hotel can be tested not through intuitive talent but rather systematically and consistently by implementing a specific empirical method.

References

Adam HA, Herrmann A, Huber F, Wricke M. (2002) Kundenzufriedenheit und Preisbereitschaft – Empirische Erkenntnisse aus der Hotelbranche. Zeitschrift für die betriebswirtschaftliche Forschung. 54: 762-778

Baetge J, Kirsch HJ, Thiele S (2004) Bilanzanalyse (2). Düsseldorf.

Bieger Th, Rüegg-Stürm J, von Rohr Th (2002) Strukturen and Ansatze einer Gestaltung von Beziehungskonfigurationen - Das Konzept Geschäftsmodell. In: Bieger Th et al. Roland Berger Strategy Consultants, Academic Network (eds) Zukünftige Geschäftsmodelle, Berlin.

Bernet B, Bieger Th (1999): Finanzierung im Tourismus, St. Gallen.

Büschgen HE, Everling O (1996): Handbuch Rating, Wiesbaden.

Diller H (2002) Zum Preisverhalten bei Dienstleistungen: Die Preiszufriedenheit verdient mehr Aufmerksamkeit. In: Hinterhuber HH, Stahl HK (eds) Erfolg durch Dienen? Innsbrucker Kolleg für Unternehmensführung, Bd. 4. Renningen.

Diller H. (1997) Preismanagement im Zeichen des Beziehungsmarketing. Die Betriebswirtschaft. 57: 749-763

Eigermann J (2001) Quantitatives Credit-Rating unter Einbeziehung qualitativer Merkmale, Frankfurt am Main.

Feri Branchen Rating Deutschland (2004): Hotels (ohne Hotels garnis), Bad Homburg.

Frei I. (1999): Expansionsstrategien in der Hotelindustrie, Hamburg.

Hamerle A., Rösch D. (2003). Risikofaktoren und Korrelationen für Bonitätsveränderungen. Zeitschrift für betriebswirtschaftliche Forschung. 55:199-222

Hinterhuber HH (1999) Die Rolle der Kundenzufriedenheit in der strategischen Unternehmensführung. In: Hinterhuber HH, Matzler K (eds.) Kundenorientierte Unternehmensführung: Kundenorientierung, Kundenzufriedenheit, Kundenbindung. Wiesbaden.

IHA Hotelverband Deutschland e.V. (2006): Hotelmarkt Deutschland 2006, Berlin.

Kaiser MO (2002) Zur Kundenzufriedenheit bei Dienstleistungen: Ein kontaktzeitpunktspezifisches Messinstrumentarium. In: Hinterhuber HH, Stahl HK (eds) Erfolg durch Dienen? Innsbrucker Kolleg für Unternehmensführung, Bd. 4, Renningen.

Keiner Th (2001) Rating für den Mittelstand, Frankfurt am Main.

Kleine DW, Anclam S (2002) Basel II und die Folgen für kleine und mittelständische Unternehmen. In: Hofmann, Gerhard (eds.) Basel II und MaK, 1. Aufl., Frankfurt.

Kley CR (2003) Mittelstands-Rating, Dissertation Universität St. Gallen.

Matzler K, Pechlaner H (2001) Management der Kundenzufriedenheit. In: Weiermaier K, Peters M, Reiger E. (eds) Vom alten zum neuen Tourismus. Innsbruck.

Matzler K, Stahl HK (2002) Was bedeutet „Wertsteigerung“ für die Führung von Dienstleistungsunternehmen. In: Hinterhuber, H. H./Stahl, H. K. (Hrsg.), Erfolg durch Dienen? Innsbrucker Kolleg für Unternehmensführung, Bd. 4, Renningen.

Pauen W (2007a) Ökonomische Bonitätsfaktoren (für das Kreditrating) von Unternehmen der Tourismus- und Hotelbranche, Dissertation, Universität Innsbruck.

Pauen W (2007b) Bonitätsbewertung von Unternehmen der Tourismusindustrie am Beispiel der Kettenhotellerie. In: Weiermair K, Peters M, Pechlaner H, Kaiser MO (eds) Unternehmertum im Tourismus – Führen mit Erneuerungen, Berlin (in print).

Paul S, Stein S, Horsch A (2002) Treiben die Banken den Mittelstand in die Krise? Zeitschrift für das gesamte Kreditwesen. 55: 578-582

Statistisches Bundesamt (2005). GENESIS-Tabelle, Ankünfte und Übernachtungen in Beherbergungsstätten. Wiesbaden.

TREUGAST Unternehmensberatung mbH (2004): Trendgutachten Hospitality 2004/2005, München

Wricke M (2000) Preistoleranz von Nachfragern. Wiesbaden.

WTO World Tourism Organization (1998): Empfehlungen zur Tourismusstatistik. In: Statistisches Bundesamt: Tourismus in Zahlen. Wiesbaden. p.12-20

Quality Rating in Hotel Community Sites

Susanne Fittkau, Axel Jockwer

1 The First Step Is Always the Hardest: Deciding on the Vacation Destination

Vacations are a scarce resource. The annual number of vacation days is limited; the anticipation of rest and relaxation for this time is nevertheless immense. And, to top it off, vacations are expensive. A family of four has to invest the equivalent of purchasing a used compact car. Would anyone really make such a purchase based solely on the dealers' sales pitch? Does not everyone know at least one person, who is a connoisseur of cars, whom they would consult before making such a purchase?

Imagine you had a very large circle of friends and acquaintances, who were all travel enthusiasts. Wouldn't it be amazing to be able to count on their experiences when making vacation plans! Every one of your friends and acquaintances surely has their own preferences and tastes; however, they are sure to provide a lot of clarity to the decision-making process. You can easily find out what someone thinks about the hotel location, what another feels about the wonderful breakfast buffet and what yet another has to tell about staff friendliness. And if someone can actually show you a selection of photos or even a short video from the vacation destination, your image of the vacation hotel will become more tangible.

Most people cannot count on such a circle of friends and acquaintances to make travel decisions. Even a large circle of friends seldom has such extensive experiences as to provide valuable information on a multitude of destinations. However, it is not actually important – for the omnipresent "new" medium, the Internet, has already rendered this deficiency significantly. The Internet has secured an invincible position as an information tool in many sectors including the tourism industry. The future trend for many industries is therefore: An increasing number of consumers are preparing and making more and more purchasing decisions over the Internet.

2 The Purchase Decision Is Made Online

The Internet is and has always served users as a medium for communication and current news. In the past years, it has nevertheless converted rapidly into an important tool to compare products and prices and to shop respectively books and/or services for example. According to the results of the renowned internet study "WWW-Benutzer-Analyse W3B" over two thirds of Internet users affirm that they consult the Internet to gather information about products and services. As part of the W3B survey, the research institute Fittkau & Maaß, located in Hamburg, surveys over 100,000 German-speaking Internet users twice a year. The survey results show that, depending on the product, up to three fourths of the users believe it is important to surf the web before making purchase. This tendency applies especially to complex, expensive products requiring consultation – including, of course vacations.

The number and diversity of product and price information available over the Internet is continually increasing. A user can now choose from a variety of different online information sources – from price comparison machines to editorial websites and even interfaces for exchanging user opinions.

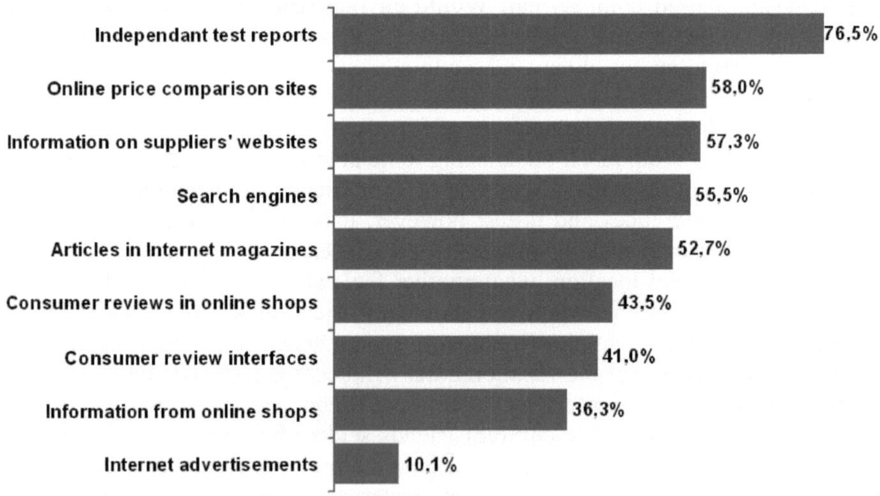

Fig. 1: Online sources of information used to make purchasing decisions; Source: W3B Survey 2007, www.fittkaumaass.de

Of all the different types of online product information, many users prefer consulting independent sources. In particular, if the information is available, users tend to consult test reports from institutions or organizations. In addition, more than half of the users rank both search engines and specialized online price comparison sites as important sources. This also applies for information distributed directly over the Internet by manufacturers or suppliers.

3 Planning Decisions in the Social Web

Even the "social web", the phenomenon of using the Internet to create social networks and communities and to exchange information and opinions, already plays an important role in regard to preparing purchase decisions. As such, numerous online platforms are already present in the Internet, where users can exchange product information and thus actively participate in the opinion making process for specific products. These platforms are either integrated on specific websites, such as online shops, or independently positioned as consumer review sites. Whichever the case, over 40% of Internet users view this user-generated content (specifically the ratings and recommendations of other users) as an important source of information when searching for products online.

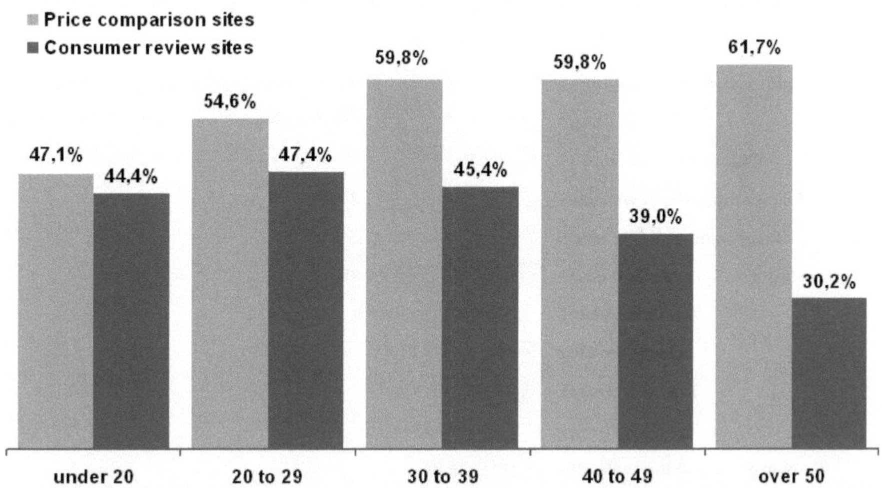

Fig. 2: Older users prefer price comparison sites, younger users prefer consumer reviews; Source: W3B Survey 2007, www.fittkaumaass.de

The preferences for the different sources of online information depend heavily on the age of the users. Whereas older users usually choose to consult information from independent sources, younger users tend to be drawn to social websites. Thus, with increasing age, users are more inclined to use independent sources of information such as reports or price comparison sites. On the other hand, interest in exchanging user reviews and opinions seems to decline as the user ages.

4 Success Story "Travels Online"

The Internet is an extremely important and interesting source of information for both suppliers and users even in the travel and tourism industry. Consequently, the topic "Travels online" has also been a key issue in the Internet market study W3B. Being a direct marketing channel, the Internet has revolutionized the travel market – and furthermore the Internet has become an indispensable part of the travel and tourism industry. In fact, "traveling and vacations" are among the most important hobbies or topics of interest for German-speaking users. Almost one of every two Internet users claims to be interested in travel. Actually, websites displaying information from travel agencies, tour operators, airlines and railway companies are among some of the most frequently consulted website categories with German content. Over one fourth of Internet users search online for flight schedules and time tables, and near to every sixth user visits travel websites on a regular basis.

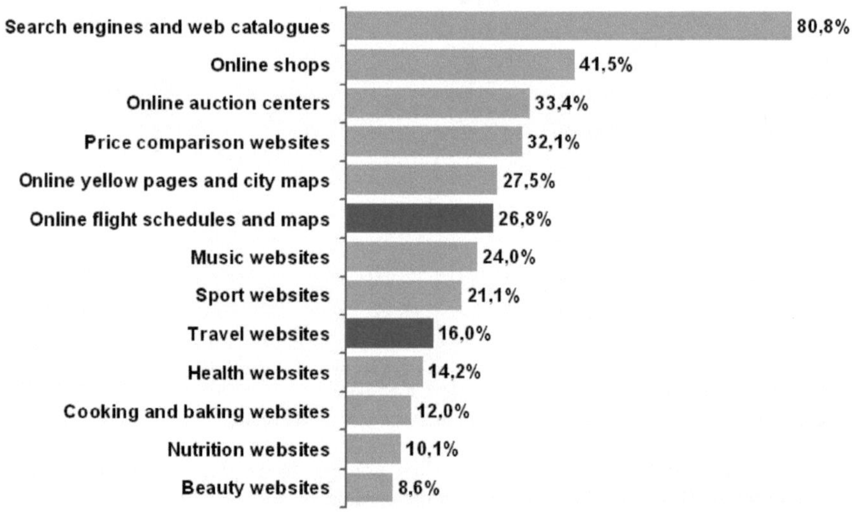

Fig. 3: Tourism and travel sites: some of the most visited website categories; Source: W3B Survey 2007, www.fittkaumaass.de

The percentage of people, who have already purchased travel products over the Internet, is increasing continually. In the spring of 2007, 40% of the German-speaking Internet users had already purchased flight or train tickets online at least once. And, according to personal accounts, no less than 30% had actually booked a vacation via Internet.

5 The Internet: A Travel Information Channel

Those who currently use the Internet usually also consult specific travel information. If one were to ask an Internet user where they looked for information before booking a trip, *online information from the Internet* is sure to be the principal source. Over two thirds of the German-speaking online users claim that the Internet is the most important resource for making informed decisions when preparing a trip. It is even more important than suggestions from friends and acquaintances (as stated by over half of the users) and clearly more important than information provided by travel agencies (as stated by just under 40% of the users).

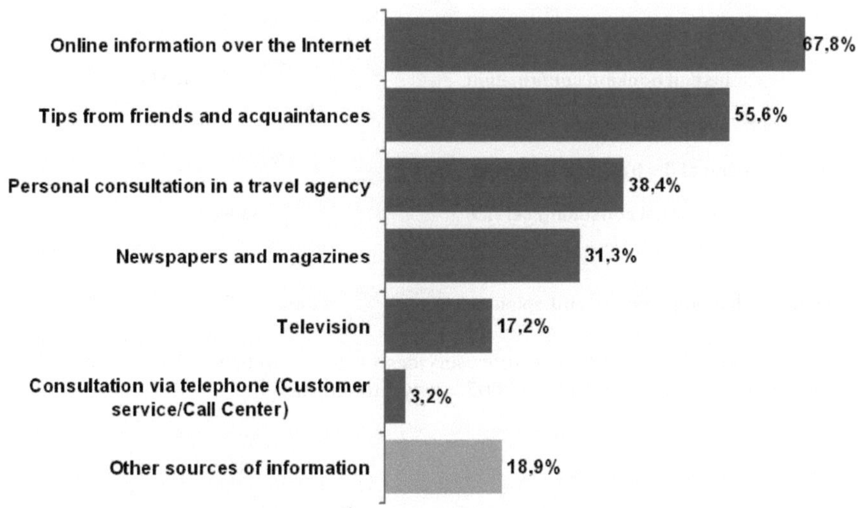

Fig. 4: The most important sources of information for booking trips and vacations; Source: W3B Survey 2007, www.fittkaumaass.de

The Internet is thus undisputedly the most important source of information for making travel decisions. For the most part, users tend to consult online information on flights and last minute trips in particular. Information on hotels, package tours and city tours, train connections and schedules follow thereafter.

Until quite recently suppliers within the travel and tourism industry were the ones distributing online information of this nature over tour operator, carrier or travel agency websites. However, the *social web* began to edge its way into the tourism industry many years ago and has nowadays become common practice. The idea is quite simple: from the customers' perspective, networking permits users to supplement the all too limited information provided by suppliers with an immense pool of private information and experiences presented by other travelers. Some suppliers make this exchange possible on their websites. After all every fifth users actually expects such service on travel websites. In the tourism industry, it is once again the younger users, under the age of 40, who like to consult consumer reviews about vacations and destinations, as a source of information and as a basis for making decisions.

Nowadays however, the social web plays a considerably larger role in the travel industry, especially concerning (independently positioned) Internet portals for tourism and hotel ratings.

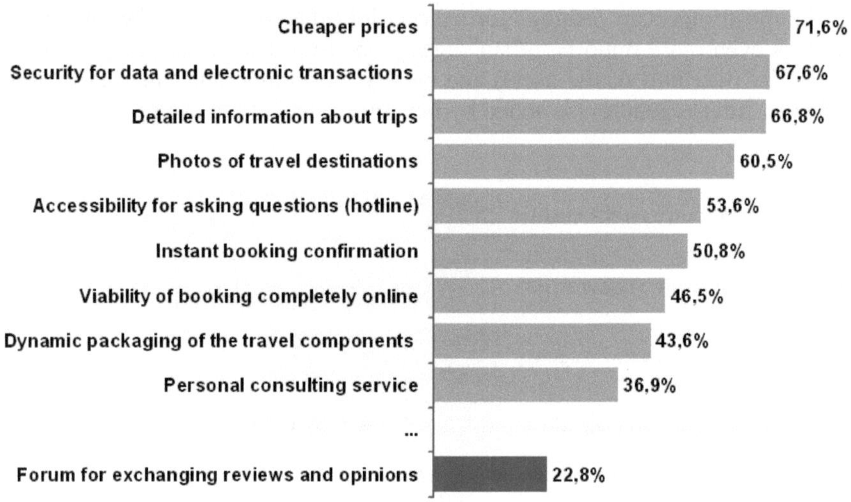

Fig. 5: The importance of certain features/services on travel websites from the users' per spective; Source: W3B Survey 2007, www.fittkaumaass.de

6 The Social Web in the Travel Industry: The Idea of Hotel Rating Portals

The concept of Internet rating portals for vacation spots or hotels is based on gathering and organizing the maximum amount of information on travel destinations, which users would otherwise have to painstakingly compile. All users can construct an informed impression with help of a hotel rating portal based on official information (i.e. catalogue information provided by organizers) and information of private origin (including user-generated content). The chances of encountering a vacation that suits the user are quite good because the potential traveler has all the resources at his fingertips and they are completely open to interpretation. The multitude of information breeds knowledge, clarity and last but not least security. In the end, most users end up avoiding surprises and disappointments.

7 The Multitude of Information in Hotel Rating Portals

How does a young couple, a family or an elderly person traveling alone rate a hotel? Do travelers praise the food or do they rather pick at their food? Do they recommend particular rooms? Do popular travel periods emerge? Will readers decide whether or not to let themselves be inspired by the written commentaries on the basis of the richness of the text, the detail of the descriptions or the quality of the orthography?

Many hobby testers even introduce themselves to other users and provide a picture and profile, thus establishing contact in order to exchange information through dialogues – either in private messenger systems or public travel forums. Planning vacations is increasingly transpiring via an information exchange network or community – together knowledge grows.

Aside from the mass of hotel ratings, countless photos depict the current vacation landscape. Regardless of whether it is at the buffet, on the beach or in the restroom – a digital camera is always in the right place at the right time. And, with increasing frequency, they also provide useful and entertaining videos.

The subjective experiences from dozens of former guests accompany an extremely large amount of official information: How do catalogues from different tour operators rate a hotel? Does the catalogue from a specific tour operator shed more light on leisure activities than those of the other operators? Do the photos in the catalogues appeal to me? Does the hotelier offer additional information? How far away from the beach is the hotel according to Google Maps? What does the system know about the weather, water temperature and medical requirements? And ultimately: What is the actual market price to stay in a particular hotel, under what conditions and with which tour operator?

The combination of a maximum amount of information with a large variety of categories and sources provides travel enthusiasts with a wide-ranging tool for self-counsel – which in all actuality is the whole purpose of hotel rating portals.

8 Online Hotel Ratings: HolidayCheck Case Study

The current market leader in the European hotel rating portals is HolidayCheck. This company provides visitors with information, of which the largest and most important segment is user-generated. Vacationer can post their opinions about a specific hotel, photos or videos of this hotel as well as the surrounding areas, tips about restaurants, points of interest or leisure activities etc.

HolidayCheck ensures that the content is correctly classified and presented and regularly performs thorough quality checks.

In order to guarantee the value of the information, HolidayCheck has to filter both: excessively positive reviews written by the hotel itself as well as attacks with criminal content. For this purpose HolidayCheck introduced a preliminary technical inspection and a manual follow-up, which is all currently being monitored by 30 team members. Not every hotel rating system is up to such a challenge; however, at HolidayCheck, it is a core element of the whole application. Although this system cannot ensure the complete absence of forgeries, the technical filters and process monitoring as well as the experience and detective instincts of the controllers keep the "hatches battened down" and maintain high quality.

The platform's operating mode regulates everything else: The amount of detailed ratings (quite a few hotels receive hundreds of ratings) and the community mechanism (opinions provoke opinions) ensure a sizable amount of valuable information, which other travelers can use to visualize and plan trips.

9 Hotel Rating Portals from the Perspective of the Organizers and Hotels

Destinations have become aware of the relevance of traveler opinions. More and more hotels are experiencing the positive and negative effects of this elevated business transparency. Two reactions can be observed: On one hand hotels oppose allegedly unjustified criticism and threaten to take legal actions – a response which does not stand a chance since the German court of law has already set precedent on two occasions, claiming freedom of speech for rating tourism services. On the other hand, there is a growing tendency among hoteliers to face public opinion, to embrace improvement measures and to take advantage of the advertising effects of positive customer reviews.

Even tour operators have started to take comments from private hotel ratings into consideration. When operators consider purchasing hotel contingent, they increasingly direct their attention to recommendations from former guests, since premium hotels with good online reputation have favorable, dynamic selling power on their own. Hotels with weak reviews sometimes have to renovate and reorganize or face being eliminated from the program.

Customer opinions influence the product and therefore serve to eventually improve the quality of the travel experience – a trend which is sure to expand.

10 Conclusions

Since the Internet and specifically the online opinion and rating portals entered the tourism market as information sources, a monopoly for providing and distributing information has become virtual non-existent. In fact, potential customers not only have the option of accessing travel information via Internet at any time and from any location in a matter of seconds, but they can even generate and post their personal inquiries or experiences. The actual reliability of this user-generated content depends of the context, complexity and presentation within each trip or hotel rating portal.

Suppliers and sales agents will have to adjust to the new times: In the future, the travel information they prepare and distribute will increasingly be enriched and enhanced – and not on their initiative but by their actual customers.

Destination Management

Improving Travel & Tourism Competitiveness

Jürgen Ringbeck, Stephan Gross, Thea Chiesa, Jennifer Blanke

1 Preface

This article is based on the results of the "Travel & Tourism Competitiveness Report 2007", which was launched and presented as part of a public panel discussion at the ITB convention in March 2007 with Booz Allen Hamilton industry research.

2 Introduction

With international tourism receipts increasing from US$ 2.1 billion in the 1950s to US$ 622.7 billion by 2006, the travel & tourism sector accounts for 10.3 percent of world GDP, creating 234 million jobs in the industry, making up 8.2 percent of total employment worldwide. Since growth is mainly fueled by international travel, countries need to be well prepared to gain a competitive advantage in the global tourism sector – which is affected more by globalization than any other industry.

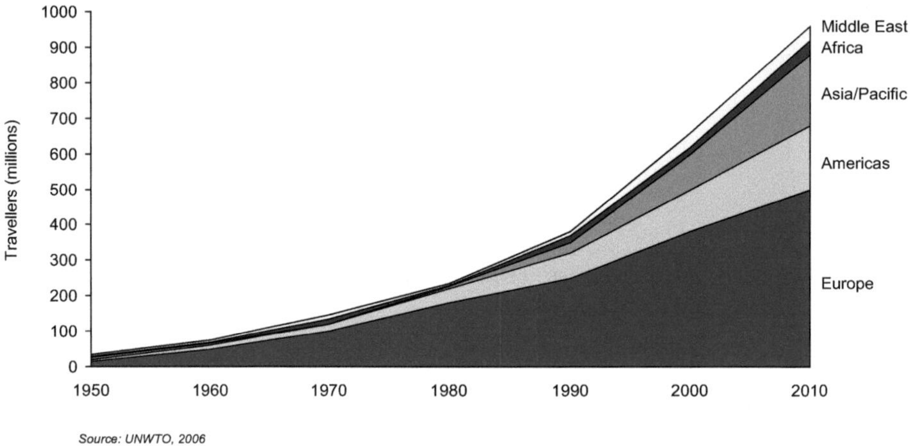

Fig. 1: Development of international travellers and global tourism income (1950 – 2006)

Travel & Tourism (T&T) is thus one of the most important economic activities internationally, and the main industry in many countries, as well as the fastest growing economic sector in terms of foreign exchange earnings and job creation. The sector is an important driver of growth and prosperity and, within developing countries, of poverty reduction. In fact, according to the World Travel & Tourism Council (WTTC), most new jobs in developing countries are created in tourism-related industries.

The sector also has important indirect positive development effects. It encourages governments to make infrastructure improvements such as better roads, electricity, telephone, and public transport networks, which, as well as facilitating tourism, improves the economy's overall development prospects and the quality of life for its residents.

The dependence of tourism on the quality of the natural environment also places it in a special position in terms of environmental sustainability. The travel & tourism industry can make a positive contribution to the quality of the natural environment by, for example, communicating the value of the natural environment to residents, by creating business incentives for environmental improvements, and through raising awareness of environmental issues and encouraging tourists to advocate environmental conservation.

With growing income and positive influence of Travel & Tourism, many countries have recognized Travel & Tourism as a critical sector that can incubate national prosperity and economic growth. But how?

3 The Global Travel & Tourism Competitiveness Report

In an effort to better understand the drivers of T&T competitiveness, the World Economic Forum embarked in the development of a tool that would provide governments and policy makers with a comprehensive understanding of the opportunities and challenges that face the national development of the T&T industry at the present time. Together with its strategic design partner Booz Allen Hamilton and other leading organizations and operators in the travel and tourism industry[1] the World Economic Forum developed the first-ever "Global Travel & Tourism Competitiveness Report", which was presented at the ITB Berlin 2007.

At the heart of the Report is the Travel & Tourism Competitiveness Index (TTCI), measuring the factors and policies that make it attractive to develop the T&T sector in different countries.

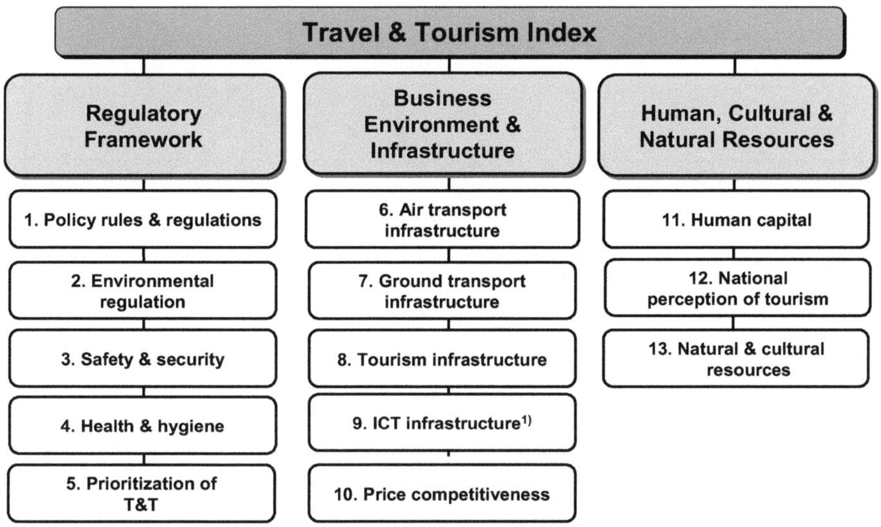

Fig. 2: Structure & Components of the Global Travel & Tourism Competitiveness Index

The structure and components of the TTCI are shown in Figure 2. The Index is based on three pillars, which are important to the competitive strength of the sector: the regulatory framework, the business environment and infrastructure, and the targeted support for existing natural, cultural and human resources. These broad pillars are further broken down into 13 sub-categories and more than 50

[1] UNWTO, WTTC, IATA, Carlson Group, Emirates Group, Visa International, Bombardier, Qatar Airways, Royal Jordanian, Silversea Cruises, and Swiss International Airlines

variables, from safety standards and environmental laws to the quality of the transportation infrastructure and staff qualifications in each country.

The importance of the various factors that make up the TTCI might vary depending on each country's stage of development. While some of the factors are a "must have" for high income economies, they might not yet be those requiring the most urgent attention for developing countries.

Safety and security, for instance, is a prerequisite for any country looking to attract foreign business and international travelers, while government investments in environmental protection and new technologies, on the other hand, might become relevant only once basic infrastructure is in place.

Comparing economies within each stage of development makes it possible to identify specific key success factors that are specifically applicable to developing or more advanced countries (see Figure 3).

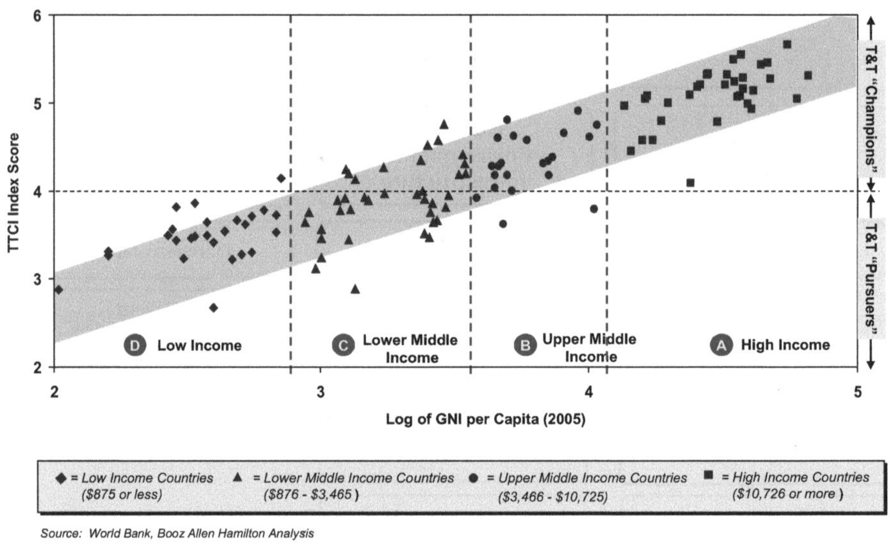

Fig. 3: TTCI Score vs. Gross National Income per Capita

Travel and tourism naturally increases as a country's economic and social welfare improves, along with the underlying supporting factors captured in the Index. That is why the TTCI naturally ranks advanced economies higher than countries at lower development stages. Taking the gross national income (GNI) per capita as an indicator, it shows that the first 27 rankings in the TTCI are all countries that belong in the 'high income' category. Only seven high income economies have been surpassed by countries in either the "upper middle income" or the "lower middle income" segments.

The TTCI score bandwidth within the four different segments shows that some countries emerge as high performers relative to their peers. Figure 4 highlights the top ten "fly-wheel" countries for each of the four segments – illustrating their overall TTCI rank and their respective scores.

'High Income' Peer Group		
Index Rank	Country	Score
1	Switzerland	5.66
2	Austria	5.54
3	Germany	5.48
4	Iceland	5.45
5	United States	5.43
6	Hong Kong	5.33
7	Canada	5.31
8	Singapore	5.31
9	Luxemburg	5.31
10	United Kingdom	5.28

'Upper Middle' Income Peer Group		
Index Rank	Country	Score
28	Estonia	4.90
29	Barbados	4.86
31	Malaysia	4.80
35	Czech Republic	4.75
37	Slovac Republic	4.68
38	Croatia	4.66
39	Mauritius	4.63
40	Hungary	4.61
41	Costa Rica	4.60
45	Chile	4.58

'Lower Middle' Income Peer Group		
Index Rank	Country	Score
34	Tunesia	4.76
43	Thailand	4.58
46	Jordan	4.52
48	Jamaica	4.41
50	Dominican Republic	4.35
54	Bulgaria	4.31
57	Morocco	4.27
58	Egypt	4.24
59	Brazil	4.20
60	Indonesia	4.20

'Low Income' Peer Group		
Index Rank	Country	Score
65	India	4.14
80	Tanzania	3.86
84	Gambia	3.81
88	Vietnam	3.78
91	Mongolia	3.72
92	Mauritania	3.71
94	Zambia	3.66
96	Cambodia	3.64
98	Kenya	3.62
101	Uganda	3.56

Fig. 4: Top 10 TTCI Rankings based on Countries' Development Stage

Based on the TTCI rankings, it is possible to sort countries by their competitive strengths and areas in which they have room for improvement. This makes the report relevant for both entrepreneurs in the tourism sector as well as private investors, as they gain better insights into markets and investments. It is an even more important tool for government agencies and associations in the travel and tourism sector, since it shows how they can shape the policy agenda and what legal framework has to be present to increase a country's competitiveness.

This is the first time such a comprehensive Index has been developed for the sector, and it will be strengthened to include additional critical elements in future iterations of the Report. Nevertheless, the"Travel & Tourism Competitiveness Report 2007" is already providing a platform for dialogue among business leaders and policy makers around the globe. The results have been presented in numerous country and regional seminars bringing together tourism ministry officials, business leaders from the sector and representatives of civil society.

4 Defining a Strategic Agenda for Travel & Tourism

Since there is an strong relationship between a country's prosperity and the competitiveness of its travel and tourism sector, we can make a general distinction between two groups of countries with different strategic positions: travel & tourism "pursuers" and "champions".

Most of the countries with a relatively low Index rating – we call them "Pursuers" – also have a prosperity level below the global average. Their task is clear: develop their travel and tourism sector further in order to increase their country's overall prosperity.

Countries with a strong Index ranking and relatively high income levels can be described as "Champions" – almost all boast an economy at an above-average development level. However "Champions" cannot merely enjoy the fruits of their success, but have to constantly defend their position. The role of tourism in economic development is set to become even more important with international competition heating up. "Champions" now have the opportunity to focus their investments on innovative, sustainable and long-term strategies and to position themselves better for the ever-changing global economy.

4.1 Travel & Tourism as an Engine for Economic Development

What exactly can the countries in the "Pursuers" group do to improve their competitiveness in tourism for the long term? It is apparent that these countries have a great deal of untapped potential. In Romania, for instance, the T&T sector currently only contributes to between one and two percent of the country's GDP, while the figure for neighboring Hungary is at five percent.

Regardless of the particular country we look at, the first order of business is to create the right framework. This task begins with observing solid safety and health standards and building a well-functioning transportation and communications infrastructure.

Key Levers for T&T Index *'Pursuers'*

- Consider T&T as a major lever to drive overall economic development
- Stimulate fast growth of the most attractive regional tourism clusters
- Provide connectivity through a world class infrastructure (ground, air, telco, utilities)
- Foster private T&T operator business through ownership/property rights protection
- Attract private capital (e.g. via public private partnerships) and open markets for foreign investors
- Ensure sustainable development by preserving natural and cultural assets
- Make the development of T&T talent a key priority in the national education programme
- Facilitate cross-border transport market deregulation (e.g. air, rail)
- Promote a local and cross-border T&T agenda attracting more inbound travel

Source: Booz Allen Hamilton

Fig. 5: Key Success Levers for Travel & Tourism "Pursuers"

A well functioning infrastructure, along with the ease of entering a country is fundamental. Without well-functioning air and ground infrastructure and an efficient network of transport service providers no region can be developed for tourism. Common fundamental quality standards also need to be implemented in hotels and restaurants to attract international tourists who are used to such standards and have come to expect them.

Rapid development of tourism often demands focusing on specific regions because comprehensive development would not be financially feasible at the outset. Governments must step in by picking the target regions and drafting an overall development concept. Failure to control growth can lead to a slump and environmental damage, problems which tourists are quick to punish by selecting alternative destinations, as many countries on the Mediterranean have had to learn first-hand experience. Focused and sustainable development of a country's cultural and natural assets, on the other hand, is vital for long-term success.

If there is one key player to give strength to the tourism sector, it is small and medium-sized businesses and entrepreneurs, for example in the hotel and restaurant industry. Private initiative is not only desirable, but also indispensable. Governments should not be afraid of giving private and international investors the freedom and incentives to invest broadly in tourism – for instance through a master development plan. Thinking in broader terms is crucial, as we are not talking about single hotel projects, but a whole regional development plan that takes into account the local residents' concerns. Public Private Partnerships, or cooperation between private companies and the public sector, are especially important in this context.

4.2 Moving from a Travel & Tourism "Pursuer" to a "Champion"

Case Study: Dubai

Over the past decade, Dubai has emerged as a textbook example how creativity, a wealth of ideas and courageous investment decisions can create an attractive travel destination and lucrative target for investments virtually out of thin air. It is fair to say that Dubai successfully copied the formula for success that transformed Las Vegas from a small desert city off the beaten track into an American tourism and entertainment center. The emirate on the Persian Gulf has managed to practically double the number of incoming international tourists over the past six years, from 3.4 to 6.5 million. Dubai is aiming for more than double that number, or 14 million visitors by 2015.

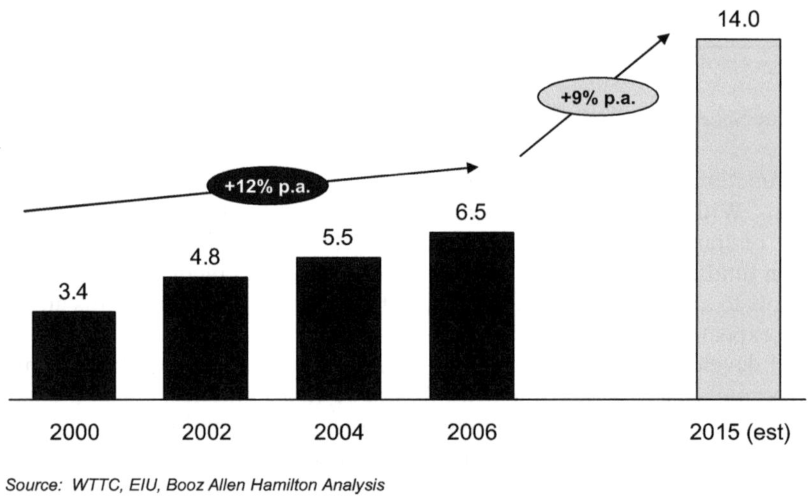

Fig. 6: Number of Incoming Tourists to Dubai (2000 – 2015 est.)

As a first move, the government deliberately subsidized foreign investments in a large number of sectors. For example, the emirate is exempt from corporate and income taxes. There are no controls on incoming flows of foreign capital, no barriers to trade, and only very modest import duties of around four percent. Dubai also boasts low wages and energy costs, creating the framework for an economic region with more than 1.3 million inhabitants, of whom 85 percent come from abroad.

But how do you induce tourists to take a vacation in a former desert outpost? The answer lies in loosening restrictions for the transportation sector. In particular, the emirate granted airlines from around the world unlimited take-off and landing rights. This deregulation was key to establishing Dubai as a global hub in worldwide tourism-related travel. Dubai leveraged its strategic position halfway between Asia and Europe and thus was able to capture its share of global traffic flows of leisure travelers. Many passengers, who are transferring in Dubai, now use the opportunity to take a short vacation on the Persian Gulf.

Dubai partnered with private companies to make massive investments in new infrastructure in order to satisfy this surge in demand. Just one example is the airport expansion for more than $4 bn. Dubai International Airport's capacity is being expanded from 22 million passengers today to 70 million, making it one of the ten biggest aviation hubs in the world. Upon its completion, the new Dubai World Central International Airport in Jebel Ali will be able to handle around 120 million passengers a year.

The lack of natural tourist attractions forced the emirate to be creative and garner international attention with outstanding architecture, sporting and entertainment events. Dubai is full of superlatives: the world's tallest skyscraper, the biggest shopping center, the biggest theme park, the biggest indoor ski arena. But it has also grown thanks to international events such as Dubai Rugby Sevens, ATP Tennis tournaments or horse and car racing. Integrating other industries into the tourism cluster ensures the emirate's prosperity. Why not combine a vacation with medical treatment? That is why there has been massive investment in medical infrastructure and training. Dubai has also gained international recognition as a prime location for the financial services and Internet-related industries.

To be sure, Dubai's example cannot simply be transferred to other countries. But the strategies it followed, the wealth of ideas and the policy approach to creating the regulatory and economic framework provide interesting pointers to all countries whose tourism industry is poised for take-off. Dubai was, much like Las Vegas early in the past century, a small desert location with few if any, attractions. Today the emirate is the role model for many industries – and an indispensable part of the global travel and tourism sector.

4.3 Travel & Tourism as a Lever to Maintain Economic Wealth

The countries in the "Champions" group possess a comparatively high level of prosperity as well as competitive strengths in their travel and tourism industry. How can these countries maintain their elevated tourism prowess and thereby defend or even expand a key part of their economic success?

Competition is increasing, and the pressure will grow to the point where domestic and international tourism flows will shift to new target destinations which today barely register on the radar. That is why it is crucial to continue promoting and supporting tourism in a targeted fashion. This also includes putting tourism at the top of policy agendas and encouraging permanent innovation in transportation and tourism services.

Key Levers for T&T Index '*Champions*'

- Consider stronger prioritization as part of the overall national policy
- Grow service quality and efficiency of the T&T infrastructure by attracting private capital and promoting public private partnerships
- Support T&T service innovation, e.g. by promoting intermodal connectivity, e-Distribution, innovative Event & Travel design ("travel as a hybrid service model"), or by stimulating seasonal demand
- Support sustainable development of T&T, e.g. by introducing traffic management systems and enforced environmental preservation
- Grow world class talent and entrepreneurs beyond Travel & Tourism (e.g. investing in educational programs, business and regulatory incentives)
- Drive forward EU cross-border transport market deregulation (e.g. air, rail)
- Strategically promote a local and regional T&T agenda attracting more inbound travel

Source: Booz Allen Hamilton

Fig. 7: Key Success Levers for T & T "Champions"

It will be crucial to keep innovating in the area of travel and tourism services in order to defend their competitive position. Tourism policy is caught between the steadily increasing demands for environmental protection and the need to use energy resources more efficiently on the one hand, and considerable bottlenecks in ground and air transport on the other.

Only an innovative transportation policy aimed at boosting the efficient use of resources such as a country's airspace can create the necessary freedom in which mass tourism can operate and thrive. Discount airlines in Europe are a good example of this. They have triggered a surge in demand that demonstrates how well-functioning market mechanisms can unleash positive forces. Additional impulses can come from "Open Skies" policies - being positive enhancers of the travel & tourism industry as they stimulate supply and enable entrepreneurs to test innovative business models.

Ultimately, tourism as a core product offering must constantly reinvent itself to live up to the rising expectations of leisure travelers who are hungry for novel vacation experiences with a local flavor. The opportunities lie in networking and connecting different regional services such as leisure, health, entertainment, education and culture so their sum exceeds the value of the parts. Each and every country can bring its respective strengths into play here. The Internet has revolutionized the distribution and sales channels for travel products and today serves as an amplifier for the best offers. It has never been easier to present innovative, compelling products and services to a global audience in a multimedia format – and to make booking them as simple as a few clicks.

4.4 Achieving High Travel & Tourism Competitiveness

Case Study of the #1 Ranked "Champion": Switzerland

With its top rank in the "TTCI" Switzerland is a strong example of how a country can continually innovate its tourism sector. In the past five years, Swiss tourism revenue increased almost ten percent and clearly outpaced the GDP growth. Switzerland has turned itself into a brand, and thus demonstrating the importance Swiss politicians and the business world assign to tourism. As a result, the country has managed to position itself in the key growth markets of the future through targeted marketing. Visitors from China are increasing by 62 percent per annum, and visitors from India, already a strong visitor base, by 27 percent.

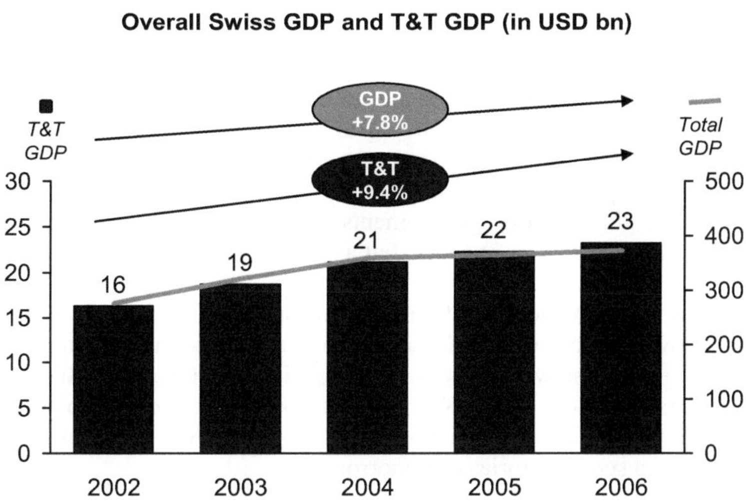

Number of incoming tourists (in millions p.a.)

2002	2003	2004	2005	2006
6.4	6.5	6.9	7.2	7.7

Source: WTTC, EIU, Booz Allen Hamilton Analysis

Fig. 8: Growth of the Swiss Tourism Sector compared to its GDP

Naturally, Switzerland can build on exceptional foundations: impressive landscapes, excellent transportation infrastructure and the potential to invest heavily in sectors that positively influence travel & tourism. These factors alone, however, cannot explain the outstanding performance of the Swiss tourism industry.

A key differentiator is the development of the whole sector, beginning with an excellent transportation infrastructure, in particular the railways. What is more, Switzerland's tourism business is geared toward sophisticated international guests and yields maximum purchasing power per visitor. Last year, roughly 10 million tourists contributed around CHF 12.9 billion to the Swiss economy, equal to almost one-tenth of the country's export revenue. Also worth mentioning is the outstanding training given to employees in the tourism sector. Switzerland is a role model when it comes to the hotel and restaurant segment, which together account for nearly half of the industry's added value.

5 Conclusion

Globalization is forcing every country - large or small, rich or poor - to compete with one another for the benefits of strong business travel and tourism. That means governments need to create an environment that best prepares their economies for the opportunities arising of the T&T sector. In an era where business travelers and tourists alike have many choices for travel and investment, every country must strengthen their Travel & Tourism sectors rapidly and intelligently, addressing the real needs of their domestic economies.

There is a wealth of ideas and approaches to help drive tourism down more innovative paths – both for countries looking to raise their game and established players. To remain competitive in the long run, each country needs to evaluate its competitive position, identify its advantages and convert them into a unique selling proposition. In this process, governments and regulators need to recognize travel and tourism not just as an industry, but as a concept that cuts across various sectors.

Dubai and Las Vegas, which both turned an inhospitable location into a shining success story show how low-valued deserts can be turned into thriving tourist destinations. Any region in the world has similar potential if it grasps potential opportunities with courage and creativity. New products and services can shape how we spend our leisure time, and regions that become thought-leaders with their innovative ideas will be the success stories of tomorrow.

European Destination Management: Challenges for Product and Brand Management

Harald Pechlaner, Petra Hedorfer, Norbert Tödter

1 Introduction

There are many reasons why European destination management has become a central figure in the tourism sector. Just like in other industries, increased competition is provoking decreasing capacity utilization, overcapacity and cost pressures, especially for traditional destinations (Ullmann 2000, p.51). In fact, these destinations have to compete with new destinations, such as those which have entered the market as a result of eastern enlargement; traditional travel flows are changing (Walder 2007, p.23) The hallmarks of globalized competition include rising mobility, travel experiences and guest requests as well as exotic features and the ability to reach far lying destinations quickly and at more and more reasonable prices (Trasser 2006, p.223). Given that the scope of tourism organizations is expanding, harsh competition for guest travel areas is rampant among providers. Marketing and product development budgets are of course also subject to competition as well as the question who will ultimately present themselves as the most competitive on the market. Therefore, traditional destinations need to rethink their positioning and develop innovative concepts in order to counteract both the declining market shares and the pressure to develop new products. The entirety of the innovations in and around the tourism industry defines a destination's success in regard to the relative competitive position on the global tourism market (Tschurtschenthaler 2005).

2 Destination Management

A destination is defined as a competitive geographical area, in which guests or target groups encounter a variety of products and services, which they consider important for their stay (Bieger 2002; Pechlaner 2000a, 2000b). These products and services may include the reception, accommodations, room service, entertainment, etc. In principle, the guest decides what appeals to him/her and thus what a destination is (Pechlaner 1999, p.125). As a result, the areas used by guests overlap, and even small destinations can be quite successful internationally. Therefore, it would be misleading to claim that a destination has to be as large as possible to be attractive and that smaller destinations do not stand a chance. As long as guests and target groups view a destination as a potential region to visit and experience, it has an entitlement to exist. A large variety of dimensions and attributes determine this perception of a destination (Page, Connell 2006). Destinations can be considered on several levels: local, regional, national, continental or global: destinations can be defined in respect to their dependency on the guests' traveling distance, target groups or markets (Bieger, 2002). These destinations can be a continent, a specific location or an attraction point. The limits of regions and destinations are intrinsically associated with the scope of the guests' activities. The further away the source market is from the traveler's destination, the larger the guest's scope of activities becomes – guests tend to take advantage of their (longer) vacation to visit a larger variety of attractions and destinations (Frömbling 1993, p.111ff.).

Destination management means offering competitive brands and products, thus consistently orienting the regions (and tourism organizations) to guest requirements. Trasser (2006) emphasizes that the value of tourism brand management is minor and that consumers practically never associate a brand with a destination. Brands, attractions, attraction points, etc. have to be interlinked and core services such as board and lodging should have adequate package marketing. In each destination, the service providers' proximity (in regard to distance and content) to competitive sectors (i.e. the clothing industry, entertainment facilities), to other service sectors (i.e. banks) as well as other trades (i.e. hand crafts) is an important driver for gaining a competitive edge in the tourism industry as well as in tourist destinations (Pechlaner 2003, p.27). A clear focus on products, markets, and business models is the prerequisite for efficient and effective Business Development.

3 From Destination Management to Destination Governance

Governance methods associate various control systems and describe the collaboration of new regulatory elements for different levels (in regard to regions). The political level also has to be taken into consideration when discussing which brands should be offered in which markets. And, when setting the priorities, either regional/local or the European level has to be selected. There are three structure models (Pechlaner, Raich 2005, p.229):

1. Market
   The best destinations prevail independently in the competitive market. They rely on a large degree of flexibility and independent decisions.

2. Hierarchy
   Generating transparent hierarchy models with a detailed classification of tasks and skills. This leads to increased formality and interdependence.

3. Network
   An intermediate amount of flexibility permits dependence but creates a balance between 'market' and 'hierarchy'.

Governance costs are transaction costs and arise from coordinating markets and businesses, and from organizing and managing a co-operation respectively (Picot 1982). The main issue here is whether it is reasonable to conduct service processes on the market or to coordinate these processes with a hierarchical system (Pechlaner, Margreiter 2002, p.196). For businesses with infrequent transactions, flexible transactions on the market are more favorable because a hierarchy or a strict organization would simply be uneconomical in such conditions. In contrast, businesses with a high transaction volume and/or highly specific transactions do better with a hierarchy model which allows them to make use of the potential for economies of scale. To be more precise, a formal and hierarchical organization is advisable when the complexity of an organization and its environment increases or when informal guidelines hinder its efficiency (Pechlaner, Margreiter 2002, p.197). In such a case, the third model 'Network' presents an acceptable middle road.

The lower the transaction costs are, the more likely the cooperation is to succeed: high transaction costs represent barriers for destination development and can provoke market failure (Pechlaner 2003, p.51). These transaction costs increase with an increasing number of partners, with less transparency between the alternatives, when there is a high level of uncertainty regarding the development of the general framework and when there is little trust between the potential partners. (Bieger, Weibel 1998). Limited capacity for information processing and communication problems between the partners in a co-operation can create a large margin for opportunistic behavior within the co-operation and utilization of dependencies (Williamson 1996). Each member operates according to a certain opportunism which implies maximizing the self-benefit and personal interests.

Traditional destinations are aware of the importance of working in co-operations in order to improve market positions and to operate jointly in innovative concepts. This requires that the countries and regions work together to create transnational travel zones for guests and target groups. Foster (1985) demonstrated this notion in respect to product development for small catchment areas, which require cooperation between the local and/or regional destination organizations in order to achieve the requirements for marketability. Since co-operations internalize positive external effects, the number of co-operations within a region becomes a competitive factor (Smeral 1997, p.109). Globalization and the resulting market dynamics are therefore additional factors, increasing the awareness for co-operations (Go, Hedges 1995).

4 Governance in the European Competitive Entity

There are two perspectives towards the possibility of creating a European competitive entity. The first views Europe as a competitive entity as stipulated on the regional and national level, which is in direct competition with other continents and travel destinations. Development in Europe has to be consciously channeled towards increasing its attractiveness on both regional and national level. The second perspective sees Europe as a destination with a coordinated governance policy capable of stimulating development in a political multi-level system with different nations and regions. Hence, the question arises as to whether a European competitive unit should rely on governance or directed services in which the countries and regions are consolidated in cooperative marketing activities or whether consistent tourism development should be encouraged or governed. Whatever the case may be, a key issue is whether or not the European Travel Commission (ETC) is capable of creating a functional classification of the European geographical area and influencing tourism development in their function as a global marketing and product development unit for Europe.

Looking at co-operations in Europe, it seems that the main issues are establishing strategic policies for sustainable tourism development, networking various players on a European level as well as implementing control media such as financial incentives. Does this mean that Europe will be a destination in the future, which can fulfill the criteria for a competitive entity? Or will Europe continue to offer services in the currently operational destinations on national, regional and local levels? Is product and market pooling the path to take, offering specific products for specific markets on national, regional and local levels? These would be the only products fulfilling the criteria for competitive units. This approach emphasizes both spatial and product specific levels. Are there products on each level which fulfill the criteria for destination management and thus serve to identify a definite direction for brand and marketing management? Which core competences, core products and attractions fulfill the competitive criteria and which can be easily conveyed to consumers and tourism organizations?

Examining these aspects on an organizational level generates three conceivable scenarios:

- **Scenario 1**: The highest European destination management authority sets approximate specifications for the core products, which the national organizations then have to prepare as specific products and services and national or regional organizations have to market (top-down approach).
- **Scenario 2**: The European destination management authority conveys short official communications and identifies the core competences and products which regional tourism organizations use as orientation in order to create competitive packages.

- **Scenario 3**: National and regional tourism organizations define the basis upon which a European tourism destination organization, for example, should distribute the products developed among the lower authorities (bottom-up approach).

However, all decisions should be taken in light of the guest perspective and not solely in regard to the (administrative) level.

At this juncture, the theory on transaction costs can be applied to a European case study. As a result, the following cost schema arises, classified according to Williamson's *ex ante* and *ex post* costs (1975):[1]

- ex ante (before transaction)
  - Information processing costs: searching for information on potential partners
  - Initial costs: contacting potential partners
  - Negotiation and/or agreement costs: (time) consuming negotiations for a partnership and intense negotiations before reaching a performance agreement
- ex post (after transaction)
  - Control costs: establishing compliance with price-performance ratio and project deadlines
  - Adaptation costs: costs incurred while adjusting to new terms and conditions for the duration of the partnership

Due to the international character of many co-operations, cultural differences should not be overlooked. Different attitudes towards objectives, values, implementation procedures and skills can often result in misunderstandings and communication problems. The different languages also present certain challenges, which prolong processes and increase transaction costs. As mentioned earlier, transaction frequency also plays an important role: in the aforementioned scenarios, operations occur in a very complex environment, making a more formal and hierarchical organization a likely possibility. However, businesses have to ensure that these hierarchies do not trigger lengthy transition or administrative expenses – as this would once again increase the transactions costs and reduce the potential for economies of scale. Another possibility would be to equate Europe with other destinations and use the market structure. Each destination would thus use their product to address a different target group. Examples would be the target group 'Asians' that finds Bavaria attractive, whereas 'Americans' is interested in Europe as a whole and the target group 'Europeans' is attracted to individual countries such as Spain or Germany.

The top-down approach clearly defines competences and creates a hierarchical form to counter the complexity and maintain low transaction costs. In contrast, a bottom-up approach makes it possible to build strong regions and nations that rely on national and international co-operations and actively exchange knowledge, which may benefit every partner. The ETC would thus internally form an authority and act as an intermediary and information hub. As such, Europe would not be

[1] Compare with Pechlaner 2003, p.51.

perceived as a single destination but rather as a conglomeration of unique (and strong) destinations.Currently, tourism organizations in various European travel destinations are clearly operating on a de facto multi-level system (local, county, regional, national, and transnational levels); however, the fundamental coordination of management and marketing perspectives in the different levels generates large transaction costs (Pechlaner 2003, p.53). In particular, setting a range for the competences is an important aspect, tasks and skills have to be clearly defined and accurately coordinated. If these fundamental requirements are not met, the cost of incorporating these skills in the aforementioned scenarios for Europe would be massive.

5 Tourism Development

Tourism is currently one of the largest and fastest growing economic sectors in the world. According to the UNWTO (United Nation World Tourism Organization), approximately 850 million overnight trips were registered world wide in international travel alone, generating around 600 billion Euros in expenditures. In addition, a multitude of domestic trips, i.e. traveling within a specific country, take place. These benchmark figures and the growth rates over the last 10 years emphasize the value of tourism today. Since 1995, international tourism has risen over 50%, illustrating an above-average dynamic in respect to other industries. As such, international tourism is currently viewed as one of the global growth drivers and tourism as one of the leading economic sectors in the 21st century.

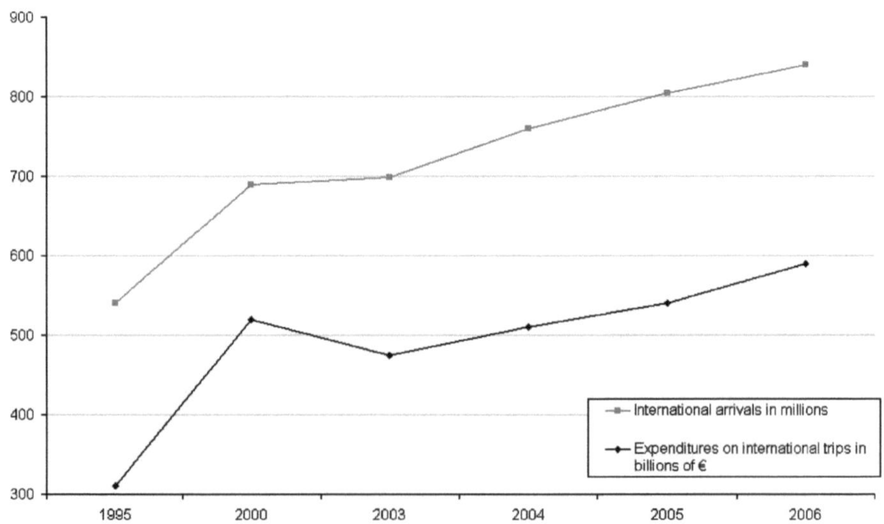

Fig. 1: The development of international tourism.
Source: Deutsche Zentrale für Tourismus e.V.

Europe is the undisputed market leader in international tourism and this situation is unlikely to change in the future: over 50% of all transnational tourism is concentrated in Europe alone; 20% correspond to the Asian/Pacific Region, 15% to North and South America, 5% to both Africa and the Middle East. Nonetheless, the UNWTO predicts that the European market share will decline to under 46% by 2020, since Asia in particular is primed to become a more dynamic destination over the next 15 years with an average growth rate of 3%.

Table 1: Growth rates and market share in international tourism

	Arrivals[a] 1995	Arrivals[a] 2010	Arrivals[a] 2020	Annual Growth % 1995–2020	Market Share 1995	Market Share 2020
Europe	338.4	527.3	717.0	3.0	59.8	45.9
America	108.9	190.4	282.3	3.9	19.3	18.1
East Asia / Pacific	81.4	195.2	397.2	6.5	14.1	25.4
Africa	20.2	47.0	77.3	5.5	3.6	5.0
Middle East	12.4	35.9	68.5	7.1	2.2	4.4
Southern Asia	4.2	10.6	18.8	6.2	0.7	1.2
World	565.4	1006.4	1561.1	4.1	100.0	100.0

[a] in millions
Source: Deutsche Zentrale für Tourismus e.V.

The 'Big Players' for destinations within Europe include the market leader Spain (over 50 million international tourists annually), followed by France, Germany, Italy, Great Britain, and Austria. This data is based on all international overnight trips including visits to friends and family. In respect to the European travel segment, vacation trips (60%) are clearly dominant, followed by other private trips (18%) and business trips (16%). Within the vacation segment, the Sun&Beach vacation is the most popular variant with a 29% market share of all international vacation trips, followed by tours (19%) and urban tourism (18%).

6 Europe – The Customer Perspective

Europe offers a large variety and diversity within a relatively small area, increasing and intensifying its tourist appeal as well as the travel expectations and sensations. Europe's product variety provides a large selection of vacations: beach vacations at the Mediterranean Sea, at numerous lakes or for a completely different experience at the North Sea, cultural and nature-oriented tours, urban tourism in world renowned metropolitan areas as well as mountaineering adventures and nature hikes in the Alps, in summer or winter. Mega-events such as the Olympic

Games, Football World Championship or cultural events (Rembrandt's 400th birthday, Mozart's 250th birthday) are typical for Europe and also contribute to increasing the tourism appeal. With this variety of tourism opportunities and vacation forms, Europe addresses a very large and diverse target group throughout the world. In short, Europe has an optimal tourism product for every target group.

Another aspect which positively effects Europe's competitive position is the area-wide (not partial!), highly developed and qualitatively first class tourism infrastructure. The air and ground transport systems and routes as well as the diverse accommodation/hotel products are just one part of it. Aspects such as security, hygiene or health care also play their part in attracting tourists (Page, Connell 2006, p.49). However, non-European competitors are already well on their way to closing this gap – providing an attractive infrastructure for tourists, particularly in the hotel industry. Therefore, it is essential that Europe takes an active stance in continual development and improvement in regard to the tourism product in order to remain competitive in the future.

In overseas markets, it is reasonable to appear with one brand 'Europe' since creating different brands for each country would be too complex. Therefore, different national tourism boards created the website www.visiteurope.com, where Europe presents itself as a single destination.

The Euro, adopted as currency in 13 partner countries (status: 2007), has made traveling in Europe much simpler and has lead to better price transparency in the market: consumers can now compare the price of individual products or services in other countries and the fees for currency exchange disappear. These features, along with the Schengen Agreement, which has been instated in 15 European countries (status: 2007), strengthen Europe's competitive position. The open border policy makes traveling easier, permitting Americans for example to explore Europe on a single visa. Another decisive advantage for European tourism lies in the fact that Europe is the largest source market for international travel. In fact, an above-average percentage of these international travelers stay in Europe, creating a large intracontinental travel volume.

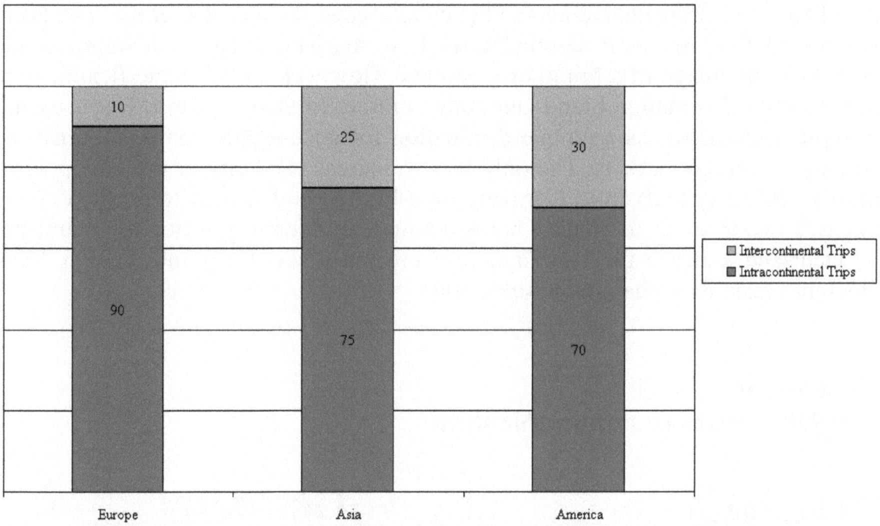

Fig. 2: European travel patterns for cross-border travel.
Source: Deutsche Zentrale für Tourismus e.V.

Furthermore, Europe holds a strong role on the demand side within the international travel market. In fact, six European markets rank among the top ten source markets for the largest demand in international travel. Germany leads the ranking, with an annual figure of around 75 million international trips, lending weight to its position as the strongest source market for international travel. Nevertheless, it is essential to bear in mind that the quality and quantity of non-European competition is increasing. Consequently, in order to ensure Europe's continued market position in the future, its product has to implement continuous improvements and innovative marketing.

Further on, the activities of the national tourism organizations with their brand and product development activities are decisive for a strong competitiveness of the destination Europe. They promote the development of core competences and core products – favored by the competition between the nations and regions within the destination system Europe. The destination Germany must also cope with the challenges at least in two areas: first, a balance must be found between the competitive brands and products of the organizations of the federal states and the Deutsche Zentrale für Tourismus e.V.; second, it is essential to continuously discuss and decide what Germany can contribute to the destination Europe in terms of products and brands.

In comparison to the international demand for trips to Germany during the EXPO 2000, Germany suffered a large downturn in 2001 as an aftermath of 9/11, greater that the world or European average: in particular, travel demands from the USA, the UK and Japan decreased substantially. In 2003, Germany started recovering considerable international demand. In 2006 with the FIFA World Championship, the international demand for trips to Germany actually increased twice as quickly as the European average and even exceeded the world average. The Football World Championship demonstrates how touristic events can improve and strengthen the image of a brand or a country. However, the strategic flexibility in managing the destination brand Germany is rather limited. In short, Germany has to improve its image as a unique destination for leisure and business tourism by catering to foreign markets. The only way to increase and ensure growth opportunity for incoming tourism to Germany on a limited budget was to create a clever image. Thus, in order to attain a homogenous brand identity in foreign countries, the Deutsche Zentrale für Tourismus e.V created a new Corporate Design 2006, which has been its global image since 2007.

Fig. 3: Corporate Design Germany.
Source: Deutsche Zentrale für Tourismus e.V.

7 Conclusions and Forecasts

The tourism sector will continue to show increasing growth rates in the future with Europe as the market leader in cross-border tourism. Over the next several years though, Europe will have to defend its market position against Asia and other attractive target markets. Europe does however hold the advantage with its superb infrastructure and previous expertise. It is nevertheless necessary to establish a European cooperation network which benefits all of the partners, in reference to new and traditional destinations, and adhere to product and brand management to which all participating members relate. The theory on transaction costs is a helpful approach for developing possible scenarios for European destination management. We have to deliberate which approach provides better opportunities for Europe.

In short, we have to determine whether European authorities adopt a top-down approach, based on hierarchical levels with a clear distribution of competences; whether Europe should compete with other countries and regions as a another equivalent destination (market structure); or whether a bottom-up approach should be implemented in which unique (strong) destinations actively participate in discussion and operate in cross-border co-operations, where the ETC could act as a possible internal authority. The fact that travelers view Europe less as a single destination but rather divide it into four regions (Northern, Eastern, Southern and Western Europe) confirms again the paramount importance that Europe examines the various alternatives and implements destination management accordingly.

References

Bieger, Th./Weibel, C. (1998) Möglichkeiten und Grenzen des kooperativen Tourismusmarketing – Schaffung von Tourismussystemen als Strategien gegen destinationsähnliche Konkurrenzprodukte, Destination Marketing (Ed. Keller, P.), Reports 48. Congress AIEST 1998, p. 167 and p. 192

Bieger, Th. (2002) Management von Destinationen, 5.A., Oldenbourg, München, Wien

Deutsche Zentrale für Tourismus e.V. (2007) Available at: http://www.dzt.de/

Foster, D. (1985) Travel and Tourism Management, p.301

Frömbling, S. (1993) Zielgruppenmarketing im Fremdenverkehr von Regionen, Frankfurt a.M.

Go F./Hedges, A., (1995) Strategic alliances, in: Tourism Marketing and Management (Ed. Witt, St./Moutinho, L.)

Keller, P. (1996) Globalisierung und Tourismus: Ein faszinierendes Forschungsthema, in: Keller, P. (Hrsg.): Globalisierung und Tourismus, Reports 46th AIEST-Congress, St. Gallen: AIEST, p. 33 – 43

Page, S.J./Connell, J. (2006) Tourism: A Modern Synthesis, 2nd edn., Thomson Learning UK

Pechlaner, H. (1999) Welche Zukunft für eine Destination Alpen? Herausforderungen bei der alpinen länderübergreifenden Kooperation, in: Fuchs, M./Peters, M./Pikkemaat, B./Reiger, E. (Hrsg.): Tourismus in den Alpen – Internationale Beiträge aus Forschung und Praxis, Studia-Verlag, Innsbruck, p.123 – 139

Pechlaner, H. (2000a) Tourismusorganisationen und Destinationen im Verbund, in: Fontanari, M./Scherhag, K. (Hrsg.): Wettbewerb der Destinationen: Erfahrungen – Konzepte – Visionen, Gabler, Wiesbaden, p.27 – 40

Pechlaner, H. (2000b) Managing tourist destinations, in: Manente, M./Cerato, M. (Hrsg.): From Destination to Destination Marketing and Management – Designing and Repositioning Tourism Products, Cafoscarina, p.9 – 13

Pechlaner, H. (2003) Tourismus-Destinationen im Wettbewerb, in: neue betriebswirtschaftliche forschung, Band 312, DUV/GWV Fachverlage GmbH, Wiesbaden

Pechlaner, H./Margreiter, J. (2002) Aufgaben eines grenzüberschreitenden Tourismusmarketings und –managements unter Berücksichtigung der Konsequenzen für die Tourismuspolitik – Das Beispiel AlpNet, in: Pechlaner, H./Weiermair, K./Laesser, C. (Hrsg.): Tourismuspolitik und Destinationsmanagement: Neue Herausforderungen und Konzepte, Verlag Paul Haupt, Bern, p. 177 – 215

Pechlaner, H./Raich, F. (2005) Vom Destination Management zur Destination Governance, in: Bieger, T./Laesser, C./Beritelli, P. (Hrsg.): Jahrbuch der Schweizerischen Tourismus-wirtschaft 2004/2005, IDT-HSG, Universität St. Gallen, p.221 – 234

Picot, A. (1982) Transaktionskostenansatz in der Organisationstheorie: Stand der Diskussion und Aussagewert, in: Die Betriebswirtschaft, p. 267 – 284

Smeral, E. (1997) Anpassungsdruck im Tourismus, in: Wirtschaftspolitische Blätter, 2/1997, p. 107 – 113

Trasser, R. (2006) (Destinations-)Marken als innovatives Verkaufsinstrument im alpinen Tourismus am Beispiel des österreichischen Bundeslandes Tirol, in: Pikkemaat, B./Peters, M./Weiermair, K. (Hrsg.): Innovationen im Tourismus – Wettbewerbsvorteile durch neue Ideen und Angebote, p.223 – 244, Schriften zu Tourismus und Freizeit, Band 6, Erich Schmidt Verlag, Berlin

Tschurtschenthaler, P. (2005) Die gesamtwirtschaftliche Perspektive von touristischen Innovationen, in: Pechlaner, H./Tschurtschenthaler, P./Peters, M./Pikkemaat, B./Fuchs, M. (Hrsg.): Erfolg durch Innovation, p.3 – 23, Deutscher Universitäts-Verlag, Wiesbaden

Ullmann, S. (2000) Strategischer Wandel im Tourismus, Deutscher Universitätsverlag, Wiesbaden

Walder, B. (2007) Tourismus – Management von Innovationen, Tectum Verlag, Marburg

Williamson, O.E. (1975) Markets and Hierarchies: Analysis and Antitrust Implications: A Study in the Economics of Internal Organization, New York-London

Williamson, O.E. (1996) Transaktionskostenökonomik, 2.A., p.13 ff.

Emerging Market Segments: Religious and Medical Tourism in India

John Koldowski, Oliver Martin

1 Introduction

Ever since the introduction into commercial air service of the Boeing 747 jumbo jet, international air travel has become not only more accessible to the world's population but also generally, more affordable as well.

Together these two elements, accessibility and affordability – at least in part – drive the billions of air trips that are now taken around the world each year.

Couple that with improved land and sea access as well and it is easy see how people are now physically connected in a way never seen before in the history of mankind.

Globally, International travel has grown from a post-WWII level of around 25 million to now more than 800 million annually (UNWTO, 1995, 2007) and expectations are that this number should reach close to – if not exceed – one billion trips per annum, by the end of this decade.

As large as that volume appears however, it fades into virtual insignificance once domestic trips become factored into the travel equation; for some countries, domestic person-trips number in the billions (PATA 2007a).

The players too have changed over time. Asia for example, once regarded as an exotic regional destination, but little else, has suddenly become one of the world's fastest growing and significant sources of travellers – both international and domestic.

Not just in numbers either. Of the top 50 international source markets last year, ranked by volume of receipts generated, eleven were Asian, with Japan, China (PRC) and Korea (ROK) all making the top ten (UNWTO, 2007).

As well as a change in the principal players across the wider travel and tourism industry there has also been a strong shift in many – but not all - of the demand forces behind this modern phenomenon. A number of the prime motivators of travel have morphed and evolved over time to encompass many different drivers: from interest in new and different cultures, experiences and interactions to others with a very specific focus.

Others have been with us for as long as modern man and remain as strong now as when they first began.

Two very good examples covering both these extremes are Religious Tourism and Medical Tourism and the place where both of these exist, side-by-side as it were is of course that country of many colours and contrasts - India.

2 India's Travel Equation

Foreign arrivals to India, after plodding along with an unremarkable compound annual growth rate (CAGR) of around 2.7% per annum during the 1980s, suddenly gained a little momentum during the 1990s (CAGR of 4.2%), but then exploded into the 21st Century averaging a growth rate of 12.0% per annum between 2001 and 2006 (Figure 1).

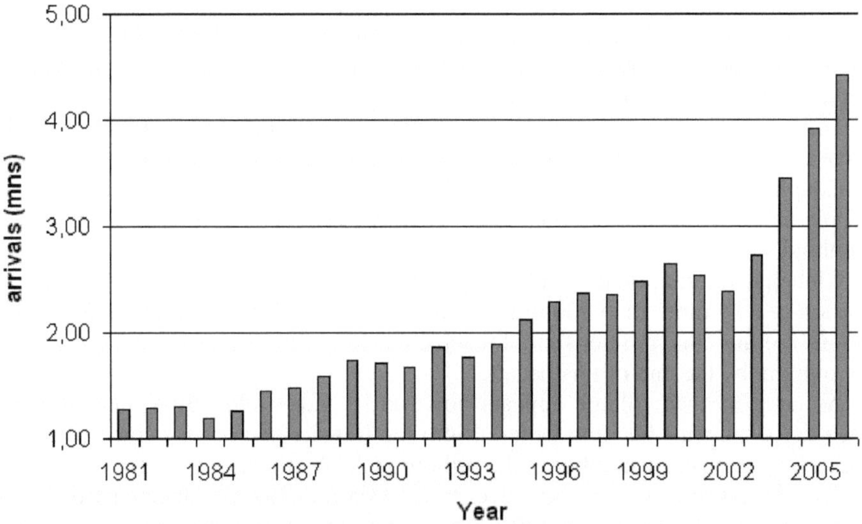

Fig. 1: Foreign arrivals to India (mns), 1981 to 2006; Source: Ministry of Tourism, Government of India (2007)

Similarly as a generating market: international departures by resident Indian nationals was relatively unremarkable through the 80's and early 90's but then began to grow dramatically, in fact almost exponentially over the last few years (Figure 2).

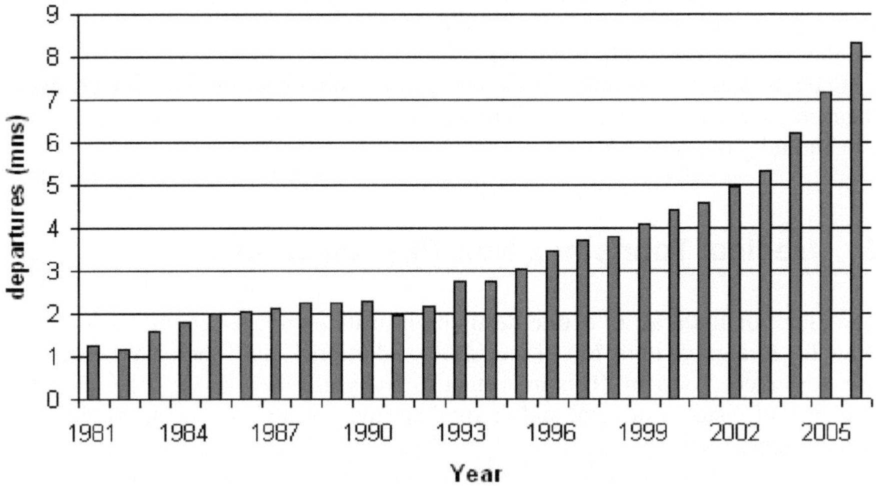

Fig. 2: Indians going abroad (mns), 1981 to 2006; Source: Ministry of Tourism, Government of India (2007)

Between 1999 - when the outbound volume first exceeded 4 million – and 2006 for example, the number of Indians heading overseas doubled to more than 8.3 million, averaging a compound annual growth rate of almost 10.7% over that period. The domestic sector has also experienced significant growth over the last decade, rising from 140 million visits in 1996 to around 461 million in 2006 (Figure 3).

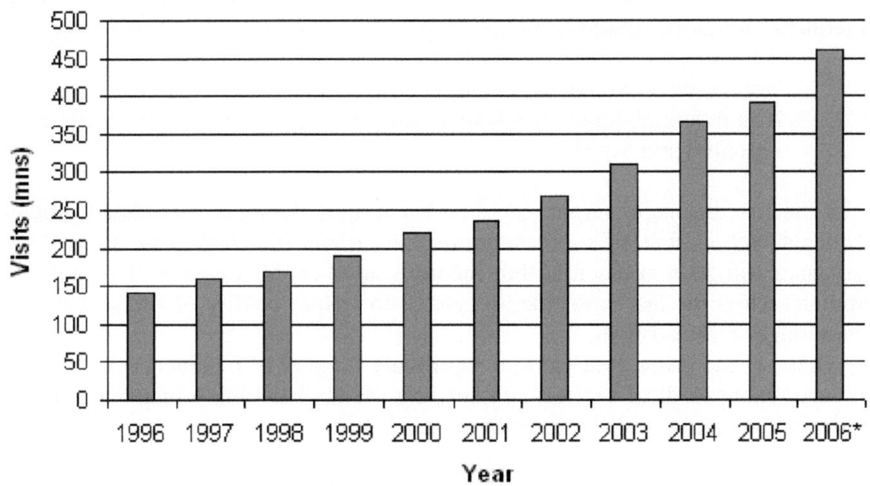

Fig. 3: Domestic tourist visits (mns) 1996 to 2006; Source: Ministry of Tourism, Government of India (2007); (*figures for 2006 are preliminary)

At the time of writing, each travel sector – inbound, outbound and domestic – was showing continued strength and resilience fuelling expectations for 2007 to be yet another record year on all three fronts.

Each of these travel dimensions however, are showing quite specific changes in their purpose mix, especially with respect to two categories at least, medical tourism and religious tourism.

3 Medical Tourism: a New Phenomenon?

The allure of India as an exotic destination in its own right continues to act as a draw for foreign visitors, but of late, a new additional 'pull' factor has entered the mix - Medical Tourism. (It is important to recognise the 'of late' rider as there is evidence of visits to mineral and thermal springs dating back to the Neolithic and Bronze ages).

As a concept, medical tourism can – at one end of what is a very long spectrum – cover those aspects of mental and physical health that are best relegated to the spas and 'wellness' centres catering to non-surgical applications. In this context medical tourism is not a new phenomenon by any means and India has been a major provider of expertise for centuries.

At the other, narrower extreme of that same spectrum however is a relatively new phenomenon – at least in current volume terms - the movement of persons offshore from their normal places of residence for orthodox medical treatment in the management of illness and/or surgery; it is this aspect of the broader 'medical tourism' sector that is considered here.

Even under this term however, there is a dilemma in-as-much as where surgery is required it may be (inter-alia):

1. Non-essential and discretionary;
2. Non-essential but recommended; and
3. Essential and urgent.

In the first instance, surgery may be for largely personal reasons and include elements such as face-lifts, breast augmentation and the like; In the second, surgery may still be non-essential but recommended on for example, psychological grounds, while the last is decidedly needed to ensure quality of life and in some cases longer-term survival.

Whatever the prime motivator, a significant number of patients seeking quality and affordable medical services are heading offshore to obtain them; furthermore evidence to date suggests that number is on the rise.

Demand for medical treatment offshore is driven by a number of factors including:

- Relative cost of procedures;
- Cost of medical insurance cover;
- Wait-times; and
- Heavy loads on existing in-country services.

With respect to cost, this is an issue not only for the individual without medical insurance coverage, but also for corporate bodies with in-house medical programmes and in some cases National Health Service agencies; all are obviously seeking the best standard of medical care but if there is a significant cost advantage as well many are prepared to go offshore to get it.

In the US for example, it is estimated that employer premiums for health care costs have jumped by as much as 73% since 2000 while employee contributions have increased by 143%. That could explain in part at least why there are an estimated 45 million people in the USA alone without medical cover and around 120 million without dental cover (Shaping Tomorrow 2006).

This makes the differences in some medical costs more than attractive to some as Table 1 illustrates.

Table 1: Relative average cost of selected medical procedures in India and the USA

Procedure Charges (US $)	US	INDIA
Heart Surgery	$30,000	$8,000
Bone marrow transplant	$250,000	$69,000
Liver transplant	$300,000	$69,000
Orthopaedic Surgery	$20,000	$6,000
Cataract Surgery	$2,000	$1,250

Source: http://www.ilsc2004.qut.edu.au

Wait-times are another issue. A 2004 study across 12 different medical specialities found that Canadians waited an average of 8.4 weeks for a General Practitioner's referral to a specialist, then waited another 9.5 weeks for treatment. Some specific surgical procedures had even longer wait times (Table 2). Furthermore, those wait times were almost double those found some ten years earlier in a similar study, so they are certainly not improving (Asian Pacific Post 2005).

Table 2: Median wait-times (in weeks) for selected surgical procedures in British Columbia, Canada (as at March 31, 2005)

Surgical procedure	Median wait time (weeks)
Endarterectomy head/neck	3.0
Cataract surgery	9.4
Gall bladder	5.1
Hip replacement	21.8
Knee replacement	28.3

Source: BC Ministry of Health, in Asian Pacific Post, 2005

This has of course presented an opportunity to a number of Canadian businesses and several have now begun offering brokerage services that arrange speedy off-shore health treatments. Similar long wait-times can be found in other countries as well; in the UK for example, it can take 1-2 years for non-emergency procedures in NHS hospitals.

Governments are also beginning to see this outsourcing opportunity as a way to decrease the demand backlog without incurring additional costs – perhaps even saving in the process; Britain's National Health Service (NHS) for example has already begun sending patients to Europe for precisely these reasons; in addition a landmark court ruling in the UK in 2006 states that the NHS will have to pay for patients to receive treatment in Europe if they face 'undue' delays in the UK.

Globally, the healthcare industry is estimated to be worth around US$3 trillion and while the value of the subset known as medical tourism is small by comparison, it is still expected to grow to US$100 billion by 2012; in India alone it could be worth more than US$2 billion annually by that same year (CII-McKinsey, 2002).

Recognising that the demand for high quality, affordable medical treatment will in all likelihood increase as the world population ages, a number of destinations in Asia – notably India, Malaysia, the Philippines, Singapore and Thailand – have all initiated dedicated, well conceived and designed programmes to tap into this growing market.

And these programmes appear to be working; currently it is estimated that India, Singapore and Thailand collectively receive more than 3 million medical tourists a year and that demand for medical services by foreigners is increasing by around 30% per annum (Alibaba.com, 2007).

Within the Singapore Tourist Board for example, there is a specific Healthcare Services Division with a mandate to develop Singapore as Asia's premier healthcare service hub for international patients. In keeping with the business focus of medical tourism this division has set a number of targets so that by 2012, Singapore can expect to host 1 million foreign patients annually, generate $2.6 billion of value-added revenue and contribute around 1% to the nation's GDP.

In India, the government also sees health care as a growth industry and with local universities producing 20,000 doctors and 30,000 nurses a year, growth in the medical tourism field would provide employment for a number of these graduates. Interestingly it is this availability of trained nursing staff and their relative affordability that gives rise to another tangible benefit of receiving medical care in India: round-the-clock personal nursing care, which is the norm rather than the exception.

Given the potential value of this travel segment in terms of revenues, investment & employment – some suggest 40 million jobs could be created by subcontracting work from the British National Health Service alone - it is not surprising that the Government of India (amongst others) is encouraging and supporting the development of medical tourism (Medical Tourism India, 2007); it is doing this through a number of initiatives including tax breaks and export incentives to participating hospitals as well as allowing long-term (i.e. one year in the first instance) medical visas to both potential patients (category 'M') and their immediate family members (category 'MX').

With 75-80 percent of healthcare services and investments in India now provided by the private sector, it is heartening to see the public sector enter a relationship that works to the advantage of both public and private sector interests.

In addition to the medical visa, the disembarkation cards required by the Bureau of Immigration for all foreigners entering India now carry an item under 'Purpose of Visit' that clearly identifies those travellers arriving for medical reasons. Clearly this will make the quantification of this travel sector all the more reliable and allow for a greater alignment with demand for services and delivery of those services.

The trickle-through effects could also be significant; the pharmaceutical industry in India for example is one of the largest in the world and even now exports products to more than 180 countries. The development of medical research facilities and expertise is also likely to rise and help make India realise its potential as an industry leader in the provision of medical services and expertise on a global scale.

4 Religious Tourism

Another aspect of India's travel and tourism industry involves the movement of people for religious and/or pilgrimage reasons; this dates back centuries.

India defines itself as a secular and multi-cultural society. Although it is predominantly Hindu, there are 144 million Muslims in India (12% of the population), 27 million Christians (2%), 24 million Sikhs (2%) and many Buddhists, Jains, and people of other religions. India is a crossroads of civilisation, being the birthplace of Buddhism, Jainism, Sikhism and Hinduism, and having been strongly influenced by Islam and Christianity.

Given this, India, for centuries has been a centre of pilgrimage and of late, a major domestic and international destination for religiously-based travel.

According to available figures, pilgrimages and trips for other religious reasons are the second most popular type of domestic trip, after social tourism, generating a 14% market share (PATA, 2007b). The Delhi based National Council for Applied Economic Research (NCAER) goes further and suggests the largest proportion of domestic trips in India are made up of religious pilgrimages (IBEF, 2004). Undertaken by both rural and urban Indians, these religious trips outnumber leisure holidays to both outdoor and urban destinations.

According to an India Brand Equity Foundation report (2004); In 2004, as many as 23 million people visited Tirupati, a temple town near the southern tip of India to catch a glimpse of a deity known as Lord Balaji. To put this into context, Tirupati's annual list of pilgrims is higher than the total number of Indian travellers visiting Mumbai, Delhi, Bangalore and Kolkata put together. In the northern state of Jammu and Kashmir over 17 million visitors in 2004, trekked uphill for 15 km to pay respects to a female goddess called Vaishno Devi (Fig. 4).

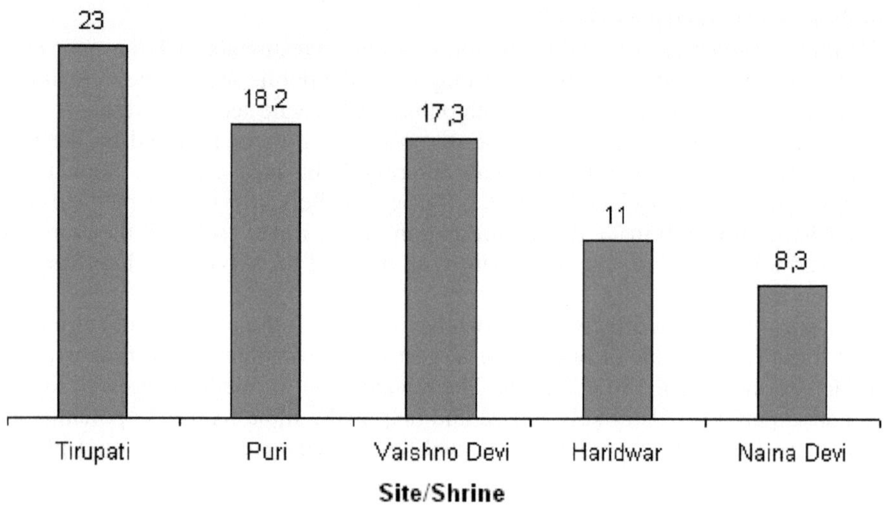

Fig. 4: Indian travellers to five main religious sites (mns), 2004; Source: National Council of Applied Economic Research, in: India Brand Equity Foundation Report (2004)

Religious tourism has not only emerged as a booming domestic market, but also a major draw for international visitors. Of all the major sectors of religious tourism, India has a captive market in the promotion of Buddhist tourism, which has huge potential for generating arrivals from Northeast and Southeast Asia markets: China, Japan, Thailand, Malaysia, Hong Kong SAR, Chinese Taipei, Korea (ROK), and Singapore. The gateways to the so-called 'Buddhist circuit' are at Patna, a city in the sate of Bihar, and Varanasi, a city in the state of Uttar Pradesh. The months of October to March form the peak season for international tourists on the Buddhist circuit. The centres with the maximum number of foreign visitors are Bodh Gaya and Sarnath, although domestic travel is highest to Sarnath.

Indian governments at the national and state level have taken several measures to develop and promote Buddhist tourism. The sites of Nalanda and Sanchi, as well as the Caves of Ajanta and Ellora, have been recognised as World Heritage Sites by the United Nations Educational, Scientific and Cultural Organization.

These and other sites of archaeological importance are maintained by the Archaeological Survey of India, a specialised governmental agency responsible for maintaining monuments.

Buddhist tourism is being extensively marketed overseas, especially through the government's overseas offices in Tokyo and Singapore. The state governments have also set up tourist reception centres at the major entry points in order to provide guidance to tourists. In the mid 2007, the Indian Railway Catering and Tourism Corporation (IRCTC) launched a new Buddhist circuit special luxury train. Covering prominent Buddhist pilgrim centres in India the train specifically targets tourists from Singapore, Thailand, Vietnam, and Sri Lanka, with a high-end luxury packages.Japanese investors are also assisting in the implementation of an integrated master plan to develop tourism infrastructure along the Buddhist circuit.

The latest phase of the Incredible India campaign, launched in 2006/2007, includes a mix of niche marketing promotions for special interest groups – including spiritual tourism. There are also plans to promote the Buddhist Circuit trail, in an effort to woo a larger number of tourists from Southeast Asia, China (PRC) and Japan.

5 Summary

India of late has become a major global player in the wider travel and tourism industry, generating more than 8 million international travellers annually and playing host to what will be close to 5 million foreign visitors by the end of 2007. That should generate around US$8 billion in foreign exchange revenues.

Add to that the domestic travel sector which currently generates around 460 million person-trips per annum and it is easy to recognise the significance of India's travel and tourism industry.

Most forecasts of tourism demand believe that the momentum gathered over the last few years will continue into the near future, especially as infrastructure catches up with growing demand, so estimates of double-digit growth in international arrivals are not surprising.

It will be interesting to watch and see what role sub-elements such as medical and religious tourism will play in the further development of India's travel and tourism industry; they will however most certainly play a role.

References

Alibaba.com, Medical tourism (2007) at: http:// resources.alibaba.com/topic/43527/Medical_Tourism_.htm

Asian Pacific Post, May 2005 'Sun, Sand and Surgery' at http://www.asianpacificpost.com/portal2/402881910674ebab010674f54cb31a55.do.html

Confederation of Indian Industry and McKinsey & Company, Health Care in India: The Road Ahead (2002)

India Brand Equity Foundation (IBEF), Indians keep the Faith: Religious Tourism Booms in India (2004)

ILSC at: http://www.ilsc2004.qut.edu.au

Medical Tourism India at: http://www.medical-tourism-india.com

Ministry of Tourism, Government of India at: http://www.tourism.gov.in

PATA, Annual Statistical Report 2006 (2007a)

PATA, Total Tourism India (2007b)

Shaping Tomorrow, Insight Newsletter: Trend Alert: Health Tourism' August 22, (2006)

UNWTO, Compendium of Tourism Statistics 1991-1995 (1995)

UNWTO, World Tourism Barometer, Volume 5 No. 2 (2007)

Marketing and Sales Management

Innovations in Market Segmentation and Customer Data Analysis

Hans Rück, Marcus Mende

1 Changes in Consumer Needs in the Travel Industry

Individualization in society is without a doubt a mega-trend for the future. Marketing activities have responded to the increasing diversity found in consumer needs, desires and longings with the concept of market segmentation, which happens to be one of the most predominant issues in marketing (Bahrmann 2002, p.86).

Table 1: Mass marketing and segmented marketing

Mass Marketing	Segmented Marketing
• Homogenous needs throughout the market	• Heterogeneous needs throughout the market, homogenous needs within market segments
• One product for one "Fundamental Market"	• A different product offer for each segment
• Competitive edge due to products with superior characteristics, pricing or advertising	• Competitive edge through differentiated products with superior fit to target group needs
• Higher profit through economies of scale in production and marketing	• Higher profit through higher margins for customized products

Source: based on Kuß/Tomczak 2003, p.63

Ever since the 1960s an increasing number of industries and sectors, starting with the consumer goods industry, have abandoned the previously powerful mass marketing in favor of segmented marketing (Table 1). The trend to address smaller segments has intensified over the last several years as a result of the advances in information technology which have made it possible to save and evaluate a mass amount of customer data, thereby fulfilling a basic necessity for one-to-one marketing and customized approaches, which symbolize the final step in this development process (Figure 1).

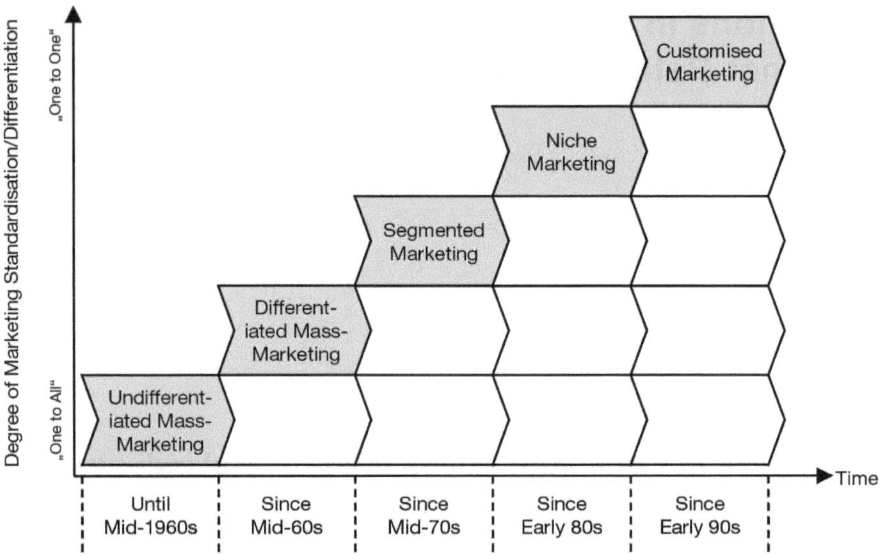

Fig. 1: The strategic trend in marketing (based on Becker 2001, p. 294)

The tourism industry has been lagging behind in this development, at least in traditional core business: classic standardized package tours are typical mass products; producers do not focus on consumer needs but on procuring and packaging transport and accommodation services in mass quantities. This concept was warranted at the outset and also relatively successful; however, its appeal declined significantly as the market became saturated. In its stead, market segments which responded to consumer demands with individualized products expanded, in particular the customized travel products, self-tailored trips and – in an attempt to put a stop to the self-service trend – dynamic trips organized by tour operators, so called customized package tours (Laepple 2003, n.n. 2007a, n.n. 2007b).

Customized package tours are going to become a very important challenge for the travel and tourism industry in the future – and a survival issue for traditional tour operators. The Internet provides consumers with the possibility to become their personal tour organizer, as Opaschowski accurately asserted: Consumers "… are turning into tour operators; self-designed dynamic trips are replacing package tours." As such, understanding and knowing your consumers and their needs better than your competitors – in short, the concept of market segmentation – becomes the decisive success factor.

2 Theoretical Principles and Problems in Market Segmentation

2.1 Definition, Goals and Strategic Principles in Market Segmentation

Market segmentation divides markets into submarkets (synonym: market segments, target groups), within which consumer behavior proves to be more homogenous than among those submarkets. Hence, specific characteristics (segmentation characteristics, classification variables) for current or potential consumers are utilized (Freter 1995, p.1803; Meffert 2000, p.181; Homburg, Krohmer 2006, p.485).

Each firm has to face the fundamental decision regarding marketing strategies of whether to offer standardized products in one market[1], aimed virtually at the "average consumer" – the unified or mass market strategy (Becker 2001, p.237) – or to offer differentiated products and services to particular customer groups or even individual customers – the segmentation strategy.

The goal of market segmentation basically consists of increasing the marketing effectiveness and marketing efficiency (where the former is an essential but not sufficient condition for the latter). The decisive factor for these two dimensions lies in a stronger identification of the target group with the corresponding product (Meffert 2000, p.183): The better the products offered are tailored to the needs of the potential customers, the higher their involvement, ensuring that the products remain attention even in times of information overload and interchangeable offers; and the higher the target group's willingness to pay, maintaining firms the opportunity of profitable growth when faced with increasingly saturated markets. Thus, ideally, greater marketing effectiveness can transform into greater efficiency and better business results.

In spite of the numerous advantages previously detailed, market segmentation is by no means always functional; it should only be implemented if the return on segmentation is positive, thus only if additional revenues arising from the segmentation are greater than the additional costs created by the segmentation.

Additional revenue from market segmentation usually arises when the consumers' willingness to pay increases in recognition of services with greater orientation to consumer needs. Customized products aimed at specific target groups are usually associated with higher prices and higher margins.

Cost savings with market segmentation usually occur if target groups are addressed precisely, thus assuring minimal wastage.

---

[1] In this context Kalmar (1971, p.105) speaks about the "fundamental market", which is the sum of all theoretical possible consumers (total market) apart from consumers, who through simple deduction, can not be considered as product users, i.e. vehicle insurance for people who do not own vehicles (Becker 2001, p.237–146, 287, 295–299).

Additional costs for market segmentation usually arise from

- increasing production costs due to increasing proliferations of options and a decreasing number of units per option (losing economies of scale) and
- increasing marketing costs arising from differentiated targeting.

The optimal size of a market segment is attained when the segmentation profit is at a maximum. A segment can also target a single consumer, as long as it is feasible and cost-effective to do so and at the same time as long as the consumer has a great amount of purchasing power and willingness to pay. However, a “segment of one” is quite difficult to achieve successfully; due to its low contact costs, the Internet clearly holds an advantage over other channels of communication.

2.2 Selection Criteria for Segmentation Characteristics

Submarkets can be formed based on a multitude of characteristics. In marketing, consumer characteristics are of elementary importance. Thus, this section will deal exclusively with consumer-based segmentation (other possible options include supply-side and competition-based segmentation).

Consumer-based segmentation always focuses on which characteristics are to be used in order to most effectively identify the diverse target groups in respect to their purchasing pattern, their accessibility, and their persuasibility. Furthermore, the division should proceed in such a manner that it is relevant, economic, and as enduring as possible for the product. Therefore the segmentation characteristics always have to meet the following requirements (Middleton 1994, p.74; Bieger 2005, p.172 et seq.; Freter 1995, p. 1807 et seq.; Meffert 2000, p.186 et seq.; Homburg, Krohmer 2006, p.486):

- *Behavior relevance*: Segmentation characteristics have to take the key drivers for consumer behavior and the respective product into consideration.

- *Selectivity*: Segmentation characteristics have to generate maximal intra-segment homogeneity while at the same time maintaining maximal inter-segment heterogeneity, specifically in relation to consumer behavior not only but also in regards to their openness to communication media in order to avoid wastage and collateral contacts. Furthermore, consumers should not be given the opportunity to gain arbitrage.[2]

- *Accessibility*: It has to be possible to address each and every one of the resulting segments.

[2] The term “arbitrage” refers to low-risk profit realization in respect to one good when different prices are set in various (sub)markets, i.e. purchasing a good on the market at a lower price and then selling the same good on the market at a higher price.

- *Feasibility*: The segmentation characteristics have to make it possible to develop specific marketing activities for the defined segments. Abstractly formulated, segmentation has to identify unique correlations between the classification variables as the independent variables and the specific marketing activities developed for this segment as the dependent variables.
- *Measurability*: The segmentation characteristics as well as the effects of segment specific marketing activities have to be measurable with market research methods (i.e. sample testing).
- *Temporal stability*: The specifications of the segmentation characteristics have to be valid on the long-term, so that the segment structure (the classification of certain consumers in certain segments) is not continually subject to change.
- *Profitability*: The resulting segments have to be yielding (financially solvent and willing to pay) and economic (which often requires that the segments are not too small). If alternative segmentation possibilities exist, the segmentation promising the highest segmentation profit should be chosen. Estimates of cost-effectiveness pose basically forecast problems because they have to be carried out before the actual segmentation takes place.

2.3 Segmentation Characteristics and Segmentation Types

Statistical methods used to identify market segments include the multivariate analysis methods such as the multidimensional scale (MDS), the multiple regression analysis, the discriminant analysis, the factor analysis and the cluster analysis (Freter 1995, p.1809).

Five segmentation types can be identified according to the segmentation characteristics (synonym: classification variables) used (Freter 1983, p.97; Freter 1995, p.1809 et seq.; Becker 2001, pp.250–293): Socio-demographic, geographic, psychographic, behavior-related and lifestyle segmentation.

2.3.1 Socio-Demographic and Geographic Segmentation

Socio-demographic segmentation combines demographic criteria (gender, age, marital status, the number of children in the household, the size of the household) and socio-economic criteria (education, vocational training, profession, income, social class, nationality, religious beliefs). These criteria are delicate: on one hand they can be easily attained, yet on the other they do not provide any direct information about a) buyers' needs and preferences or b) their purchasing patterns.

Geographic (macro-geographic) segmentation is the most traditional type of segmentation. It divides the market in regional units (such as the Nielsen Zones). Therefore, this method is counted among the demographic segmentations. As such, statistical data concerning the place of residence is fairly easy, quick and relatively cost-effective to obtain – a definite advantage. In contrast, its relevance for purchasing behavior is fairly low – a substantial disadvantage (Meffert 2000, p.191; Bahrmann 2002, p.92).

Micro-geographic segmentation is a development of macro-geographic segmentation, based on the hypothesis that people with similar lifestyles, social status and purchasing behavior tend to cluster geographically. This assumption is quite realistic, it can actually be seen on a daily basis and has even found its way into everyday language, which differentiates, not without cause, the "upscale neighborhoods" from the "blue-collar neighborhoods" etc. The significance of micro-geographic segmentation increases in correlation with detailed data; however, it is intrinsically entwined with additional costs for data collection and administration due to migration – since the population on a micro-geographic level is constantly changing.

Nonetheless, not a single method in the whole range of socio-demographic segmentation can resolve the central issue: their lack of relevance to purchasing behavior. Attitudes (to brands, products and firms) as well as communication and purchasing behavior can differ greatly between people of the same age, gender, education and income level, and within the same socio-demographic group. Thus, where segmentation is concerned, socio-demographic characteristics should only be used in combination with psychographic and behavior-related variables.

2.3.2 Psychographic Segmentation

Psychographic segmentation makes allowances for this deficit at least partially. It is based on classification characteristics such as attitudes, motives, and perceptions. In consequence, it relies on imperceptible variables, i.e. on hypothetical constructs, which require operationalization.[3] Thus, they entail greater assessment complexity than socio-demographic variables, which are relatively easy to measure.

Psychographic characteristics can be measured either on a general personality basis or related to a certain product (Meffert 2000, p.196). In the former case, the capacity to predict the actual purchasing behavior is low. In the latter, the predictability increases; however, it also requires greater data collection costs given that the influence of each segmentation variable on the product-specific purchasing behavior has to be identified. Ultimately, even product-specific classification variables are not reliable predictors for purchasing behavior – a specific attitude does not implicitly guarantee a corresponding behavior. In conclusion, psychographic segmentation lacks functionality and behavioral relevance.

[3] Operationalization means associating a theoretical construct – i.e. an imperceptible variable (e.g. "brand loyality") – according to certain rules with perceptible variables, the so called indicators (e.g. "number of different brands bought in one product category within one month") (Trommsdorff, Bleicker, Hildebrand 1975). The value of the indicators must be perceptible with the senses and in consequence measurable, and must provide reliable conclusions for the value of the variable (the theoretical construct). Operationalizations are always substitute measurements. They are never "true" or "false", rather always more or less functional.

2.3.3 Behavior-related Segmentation

Behavior-related segmentation avoids these deficiencies by making purchasing behavior itself the classification variable, such as product or brand selection, price elasticity, preferred shopping locations etc. Behavioral relevance is almost intrinsically integrated in this type of segmentation right from the start. This is the greatest strength for this segmentation – but also its greatest weakness, because behavior-related classification variables do not provide any information concerning the causes for the observed behavior, be they psychographic factors (such as attitude, motive, values) or socio-demographic (such as age, income, place of residence).

Psychographic segmentation and behavior-related segmentation are complementary; the strengths of one approach are the weaknesses of the other.

2.3.4 Lifestyle Segmentation

The classic "dilemma in market segmentation" (i.e. Meffert 2000, p.214 f.) lies in the fact that socio-demographic and geographic characteristics can be easily, quickly and cost-effectively obtained, yet are inapt as predictors for consumer behavior. In contrast, psychographic and behavior-related characteristics are well-suited as predictors for consumer behavior, yet require complicated, expensive and prolonged data collection processes.

Lifestyle segmentation is an attempt to resolve this dilemma. It is by far the most effective development in psychographic segmentation, classifying consumers according to how, generally speaking, they lead their life.

The A–I–O approach defined by Wells and Tigert, is a classic concept for lifestyle segmentation developed early on in segmented marketing (Wells, Tigert 1971, p.27; Becker 2001, p.257; Kroeber-Riel, Weinberg 2003, p.560): The letter "A" stands for activities, such as work, leisure, shopping; the letter "I" stands for interests, in relation to family, home, career, food; and the letter "O" stands for opinions, about oneself, politics, economics or nature. The A–I–O components incorporate the three most important areas for expression; in particular observable behavior (activities), emotional conduct (interests) and cognitive behavior (opinions).

Lifestyle segmentation usually classifies consumers according to several lifestyle dimensions; thus associating behavior-related aspects with psychographic and socio-demographic aspects and classifying consumers according to their similarity to "types" (statistical clusters). As such, they are often termed as "lifestyle typologies".

Lifestyle segmentation is already quite prevalent in the travel and tourism industry (Bahrmann 2007, p.97; Freyer 2005, p.188). However, there are numerous consumer typologies in use; a standard typology has yet to be agreed upon. And even if there were such a typology, it is likely not to be functional considering the multitude of possible goals.

From the practitioners' perspective, typologies are very valuable since they concisely present consumer profiles and thus make it easier to put oneself in the target group mentality. Nonetheless, it is critical to note that stereotypical segment definitions carved in stone ("Martin and Martina – the trendy bandwagon", the "Best Agers" and similar definitions) can interfere with a differentiated understanding of the market.

2.4 Current Problems in Market Segmentation

Although lifestyle research has almost completely succeeded in solving the "market segmentation dilemma" by incorporating and combining socio-demographic, psychographic and behavior-related variables, a closely related problem still lingers: linking lifestyle segments with specific address data for approaching customers.

Lifestyle segmentation permits a multidimensional description of target groups; yet, the individuals behind this description are still hard to localize – and localizing customers is a vital requirement for approaching them. As such, the "missing link" in market segmentation lies in developing database marketing[4]: The rampant development in data processing which has occurred over the last decade has made it possible first of all to build a database (data warehouses) for the massive amount of information on target persons and secondly to assess the vast amount of resulting data by intricately combining consumer characteristics.

The requirements for approaching individual customers in a targeted and more efficient manner have therefore been fulfilled through the implementation of mass customized direct marketing activities (Bruns 2007, pp.24–30).

An approach, which marks a large step towards transforming abstract lifestyle segments into tangible adress data, will now be presented.

3 The Psyma and Schober Tourism Types

3.1 An Innovative, Integrative Segmentation Approach

Schober Information Group and Psyma GmbH have jointly developed a multidimensional segmentation approach, which successfully extrapolates market research information from a random sample of persons (psychographics) to specific household addresses. Consequently, specialized, individually defined target group types can be defined for database marketing – based on reliable market research methods and approaches.

The advantages of this approach lie in the target group specific combination of psychographic, lifestyle-related and socio-demographic segmentation methods. In this respect, the missing link between market research and database marketing has been almost completely fused.

---

4 Database marketing is "marketing based on individual customer data, which have been entered in a database" (Link, Hildebrand 1994, p.107)

This multidimensional segmentation method is based on precisely structured, address-related behavioral data, as part of a standardized lifestyle survey directed at consumers since 1997. Each survey questionnaire includes approximately 160 different questions with approximately 1,500 different possible answers (characteristics). All of the survey participants give their consent to use the data for market research and advertising purposes. This lifestyle data can be classified according to the psychographic characteristics within each "customer type" (segment). In this manner, the psychographic segments can be projected onto the related address data.

The individual steps required in this segmentation method are detailed below (Figure 2).

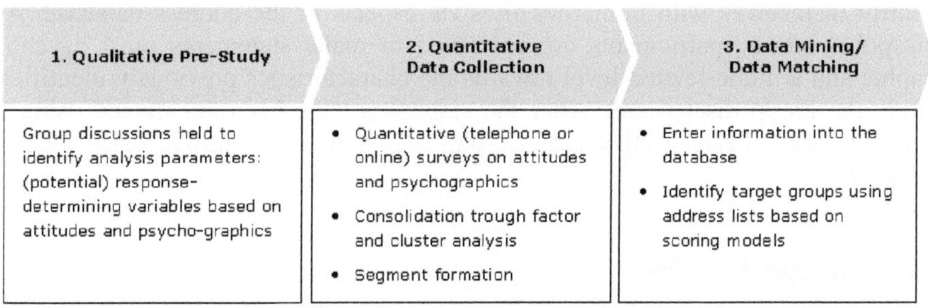

Fig. 2: Multidimensional segmentation in the tourism industry

3.2 Qualitative Pre-Study

To begin, the qualitative pre-study identifies the characteristics which are potentially useful for differentiating consumer types.

At this stage, discussions concerning the scope of behavior-related attitudes are held and the motives for these attitudes explored. These two steps facilitate the formulation of hypotheses about the respective segments in the population and generate information for approaching these segments with various marketing instruments, esp. communications (contents, reason-why, design). Qualitative group discussions play an important role at this stage since the initiated actively dynamic processes between the group participants and the different perspectives and facets update and reveal the research bases. To a large extent, these focus groups make it easier to perform a psychological analysis of consumer attitudes and motives.

Potential questions to discuss within the focus groups include:

- Which attitudes correspond to the different travel methods?
- Which spontaneous associations, likes or dislikes do consumers have?
- Which image, what specific knowledge do consumers have about the organizer and its products?
- What are the motives behind these attitudes?

- How do consumers rate the current marketing instruments, contents, arguments etc?

3.3 Quantifying the Results

In the second stage, the examined classification variables are converted into a quantitative survey format. The primary objective is to qualify available addresses with lifestyle data in regard to psychographics and attitudes.

Since database marketing requires large sample sizes, quantitative surveys include up to 10,000 complete interviews performed online and/or per CATI (computer aided telephone interviews). All of the participants in the quantitative survey identify themselves with numerous lifestyle aspects in the address database. At this point, survey participants are prompted to make statements on a psychographic and attitude-related level towards the characteristics previously identified within the group discussions. After the responses have been aggregated using a factor analysis, the respective target group segments are identified using cluster analyses.

3.4 Lifestyle Profiling

The address data attained from the survey participants are enriched with lifestyle data – via a cluster affiliation code. A profile emerges based on behavior and personality characteristics as well as product preferences and socio-demographics. Consequently, all of the defined target group clusters can be described with selective lifestyle profiles (i.e. place of residence, career, leisure and consumer behavior, product preference, interests etc).

3.5 Data Matching

In the last stage, the identified segments are incorporated in an arbitrary address list using special models, such as a household list for Germany. Thus, the available address database can be used to identify new customer potential.

First of all, in the data matching stage, every single characteristic is revised using a univariate analysis: The selectivity is verified in relation to the differentiation between target group clusters and the reference group (i.e. all survey participants). These selectivity characteristics are then associated with multivariate optimization models (scorings). The individual models usually consist of 15 to 20 characteristics, which influence the models with distinct weightings (depending on how they affect the definition of the individual target group clusters).

Using these models, each typology created by market research can be applied to address data lists; in other words, every address can be assigned an affinity to a particular target group cluster. Hence, it is possible to select addresses which represent the profile of a particular target group cluster.

In this case, the typology method is based on one of the largest household databases in Germany: It consists of around 50 million private addresses, over 200 household characteristics, with which each household can be assessed, and approximately 400 different selection criteria. This database permits marketing specialists to combine information from various subject areas: consumption issues, building data/house ratings, regional, statistical and demographic data. In addition, special mail-order data (i.e. information about the family structure, consumer spendings, the affinity to bargains or to financing issues, etc.) can also be used to identify target group addresses.

4 Case Study: The Psyma and Schober "Tourism Types"

This typology is based on a quantitative online survey, conducted by Schober Information Group and Psyma GmbH. The sample test includes 16,046 participants in a survey period ranging from 2007.01.22 to 2007.01.31.

In addition to basic attitudes and motives on travel and tourism, Psyma and Schober identified five tourism types (Fig. 4):

Type 1: Young & Fun ("get together with guaranteed action and fun"): This group's main objective is to "have a party", enjoy the night life and adventure, meet new people at the destination; they often tend to spend their vacation in clubs or hotel resorts because there are so many activities to do with other people. In fact, young travelers usually want to enjoy new experiences and do so as far away from home as possible – or at least away from home. Hence, they often choose low cost package tours, which are quick and easy to book. In spite of their high affinity to the Internet, they still prefer to consult services in travel agencies.

The "Young & Fun" type includes men and women between the ages of 18 and 29. They are single, do not (yet) have any children, are employed, students or in training programs. They have low to middle level incomes. They often have completed secondary education, and one of their hobbies is surfing in the Internet. Attending rock/pop concerts or open air festivals is also a very important leisure activity. In addition, they often like to watch music channels on television. The "Young & Fun" type represents the target group for club and all-inclusive vacations, and is also an important customer group for low cost airlines.

Type 2: Upscale Recreation Tourists ("pleasure is booked 'all inclusive'"): Upscale recreation tourists set a very high value on vacations and are even willing to spend more money on them. They are looking for pleasure and relaxation during their top-quality beach vacations and prefer accommodations in premium and superior hotels, where good food and drinks are staple items. In addition, upscale recreation tourists also like to do something for their health and fitness; finding equilibrium between relaxation and activity is important. They value service and comfort; they also prefer to consult travel agencies when booking vacations. And, they would like to have competent on-site contacts at the destination, so that do not need to worry about anything. They prefer to maintain a certain level of security, meaning that a vacation destination should not be too unfamiliar.

"Upscale Recreation Tourists" are often married 50-plus couples. Their children have already left home. They have a higher education, high school or university degrees, and are gainfully employed. They often have a high net income as freelancers, self-employed or civil servants in higher positions. They often play golf in their free time, are interested in wine and delicacies, and tend to read the "Financial Times Deutschland" and "Manager Magazine" on a regular basis. They often drive SUVs or convertibles and brand vehicles in upscale segments such as i.e. Lancia, Alfa Romeo and Mercedes. Aside from having participated in spa vacations, many "Upscale Recreation Tourists" have already booked or plan to book cruises.

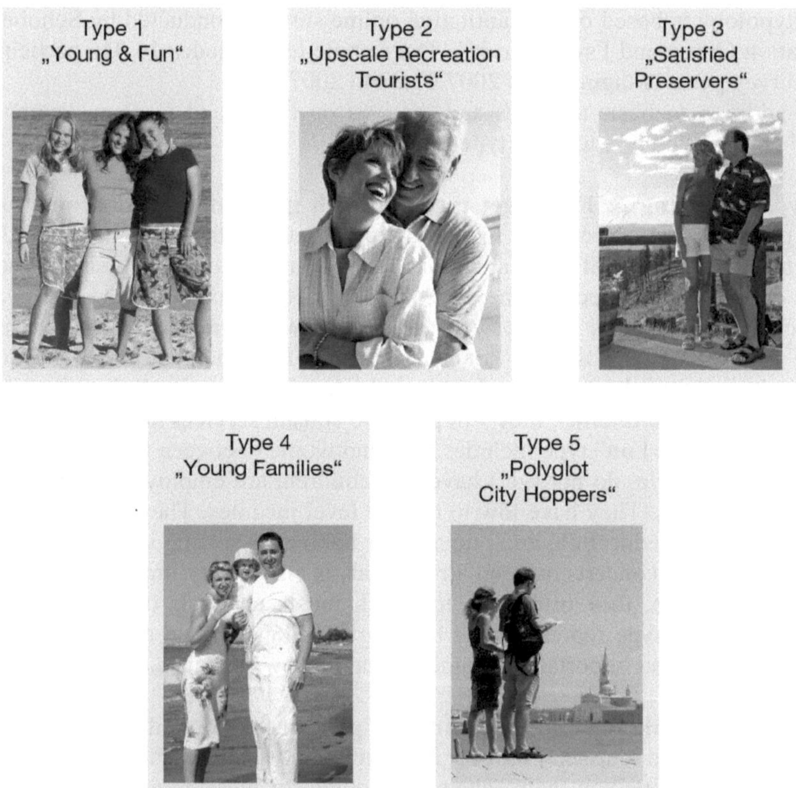

Fig. 3: Tourism types according to Psyma and Schober

Type 3: Satisfied Preservers ("vacation like in Germany with safety guarantees"): "Satisfied Preservers" often spend their summers in Germany or in other European countries, where "German is spoken". They enjoy returning to the same destination or even in the same hotel every year. Access to familiar media such as the "Bild-Zeitung" (a daily German tabloid) is important when they are on vacation. Moreover, they would like to avoid any kind of confrontation with foreign cultures and customs or social differences (poor/rich). They prefer to remain in their social environment. It is also important to avoid certain dangers such as crime or natural catastrophes. They enjoy planning their vacation close to nature, i.e. at camping sites, where they can spend as much free time with their family as possible.

"Satisfied Preservers" are often men and women between the ages of 30 and 50, who travel with family. They are married, have several children and full-time jobs, but have a below average net income. They read "Computer Bild" (a simple computer magazine) on a regular basis in their free time, drive cars from mid-level segments such as i.e. Kia or Hyundai. Many "Satisfied Preservers" own a travel trailer, are interested in motor homes and often travel to their vacation destination in passenger vehicles or motor homes. Aside from "classic" camping, these travelers also enjoy vacation parks.

Type 4: Young Families ("family focus is of utmost importance"): "Young Families" search for vacations oriented towards family and children. On-site day care facilities are especially important. This segment likes to spend as much quality family time together as possible, preferably with a relaxing beach vacation. They strive for stress-free and relaxing vacations: They want to enjoy the sun, sand and sea.

This segment includes men and women between the ages of 30 and 50, most of who are married and have one or two children. They have secondary education, are employed and have an average net income. They usually drive minivans, which do not always have to be top brands; many people drive Kia and Skoda. They are pragmatic and prefer to spend their time in vacation parks or book club or all-inclusive vacation packages.

Type 5: Polyglot City Hoppers ("traveling as a cultural experience"): They have changing destinations and travel in cities and faraway countries. They tend to organize their own trips, but also consider package tours when visiting cities. Before embarking upon a journey, they consult information about the destination in different channels such as the Internet or travel guides. In particular, they like to take in cultural activities, learn about new cultures and customs and visit points of interest. They tend to spend relaxing vacations in the wilderness as opposed to the beach.

"Polyglot City Hoppers" are usually women between the ages of 20 and 50, who are usually single but often have a partner. They usually live in one-person households, have high school or university degrees. This group often includes students, self-employed and sometimes even civil servants in higher positions, who like to read in their free time, attend open air festivals or rock/pop concerts and are interested in history, art and antiques, theater and culture. They prefer sport activities such as diving, windsurfing, snowboarding or sailing, like to look at “arte” and read intellectual media like “Geo”, “Spiegel”, “Die Zeit”, “Financial Times Deutschland” or “Süddeutsche Zeitung” on a regular basis.

Over the last twelve months, many “Polyglot City Hoppers” have taken three or more trips – in particular adventure trips, cultural trips, urban tours, educational trips, language programs, or round trips. In order to reach their destinations, they prefer to use airlines or trains. An above average percentage of these travelers have a Bahncard (frequent passenger discount card) and book only their plane and train tickets over the Internet or in a travel agency.

4.1 Address Potential for the Psyma and Schober “Tourism Types”

The address potential for these tourism types currently present in Germany are detailed in Table 2 (the database potential still overlaps for certain types).

Table 2: Database potentials for “Tourism Types”

Tourist Type	Database Potential (Addresses)	Percent of Segment
»Young & Fun«	10,656,000	18 %
»Upscale Recreation Tourists«	13,024,000	22 %
»Satisfied Preservers«	11,840,000	20 %
»Young Families«	13,616,000	23 %
»Polyglot City Hoppers«	10,064,000	17 %

Source: Psyma and Schober

5 Conclusions

The main advantage to the multidimensional segmentation approach lies in the fact that it makes the transition from market research segmentation to database marketing possible.

- The potential per segment (target group clusters) can be localized in diverse levels – from countries, counties and postal code zones to the smallest unit of a single household.
- For the first time ever, it is possible to analyze i.e. the strengths and weaknesses for the marketing potential within a catchment area of i.e. a travel agency, or even illustrate a particular competitive position.
- And, the approach also makes it easier to implement personal direct marketing for new customers in specific segments.

Thus, lifestyle segmentations can be represented in customer databases – that is, with regard to tourism and travel, the manner in which people travel and book trips. Therefore, in each customer segment, it is possible to

- define the importance of the different distribution channels (travel agencies, online, telesales, catalogue etc);
- decide which arguments should be used for particular products;
- accurately define each product mix, which has success potential; and
- calculate the individual customer durability and the probable customer lifetime value (relying on additional data).

References

Bahrmann M (2002) Die CHAID-Analyse als neue Methode der Marktsegmentierung im Tourismus: multivariate Zielgruppendifferenzierung am Beispiel Rheinland-Pfalz. Diss., Trier.

BAT Freizeit-Forschungsinstitut (2007) Deutsche Tourismusanalyse 2007. Hamburg

Becker J (2001) Marketing-Konzeption. 7th edn. Munich

Bieger T (2005) Management von Destinationen. 6th edn. Munich

Bruns J (2007) Direktmarketing. 2nd edn. Ludwigshafen 2007

Freter H (1983) Marktsegmentierung. Stuttgart

Freter H (1995) Marktsegmentierung. In: Tietz B (ed.) Handwörterbuch des Marketing. 2nd edn. Stuttgart. pp. 1802-1814

Freyer Walter (2001) Tourismus: Einführung in die Fremdenverkehrsökonomie. 7th edn. Munich

Homburg C, Krohmer H (2006) Marketingmanagement: Strategie – Instrumente –Umsetzung – Unternehmensführung. 2nd edn. Wiesbaden

Kalmar REJ (1971) Ein Stufenprogramm für Marktsegmente. Marketing Journal, 4, pp. 105–107

Kroeber-Riel W, Weinberg P (2003) Konsumentenverhalten. 8th edn. Munich

Kuß A, Tomczak T (2002) Marketingplanung: Einführung in die marktorientierte Unternehmens- und Geschäftsfeldplanung. 3rd edn. Wiesbaden

Laepple K (2003) Am Ende der Krise – wohin steuert die Tourismuswirtschaft? Grundsatzreferat, gehalten beim Sächsischen Tourismustag am 20. November 2003 in Leipzig

Link J, Hildebrand VG (1994) Database Marketing und Computer Aided Selling. Marketing ZFP. 15, pp. 107–120

Meffert H (2000) Marketing. Grundlagen marktorientierter Unternehmensführung: Konzepte – Instrumente – Praxisbeispiele. 9th edn. Wiesbaden.

Mende M, Wachter B (2007) Lifestyle Segmentation Research: Identifikation von Zielgruppen für die Neukundengewinnung im Direktmarketing. In: Herter M, Mühlbauer KH (eds) Handbuch Geomarketing, 1st edn. Heidelberg, pp. 267–270

Middleton VTC (1994) Marketing in Travel and Tourism. 2nd edn. Oxford.

Noack HC (2006) Die Reisebranche erfindet sich neu: Reaktion auf das sich ändernde Kundenverhalten und den Einfluss des Internet. Frankfurter Allgemeine Zeitung, 18. September 2006, 217, p. 23

N.N. (2007a) Wie werden wir in Zukunft reisen? Die Szenarien. Die Welt Online, 23. Februar 2007. View article at www.welt.de/reise/article732343/Wie_werden_wir_in_Zukunft_reisen_Die_Szenarien .html vom 23. September 2007

N.N. (2007b) Abschied von der Pauschalreise. Die Welt Online, 4. März 2007. View at www.welt.de/wams_print/article745021/Abschied_von_der_Pauschalreise.html from 23 September 2007

Trommsdorff V, Bleicker U, Hildebrandt L (1980) Nutzen und Einstellung Wirtschaftswissenschaftliches Studium, 9, pp. 269–276

Wells WT, Tigert DJ (1971) Activities, Interests, and Opinions. Journal of Advertising Research. 11, pp. 27–35

Customer Segmentation Traveller Types and Their Needs in 2020

Dan Greaves

This chapter will outline the findings of a report commissioned by Amadeus, the global leader in technology and distribution solutions for the travel and tourism industry, and produced by the Henley Centre HeadlightVision (HCHLV), a strategic futures and marketing consultancy, entitled *Future Traveller Tribes 2020.*

The objective of the report was to understand potential new customer segments that will emerge over the next 10-15 years and how the airline industry can meet their needs through better innovative thinking and the better use of technology. The report was not intended to be a definite view of the future but instead to stimulate discussion about what possibilities lie ahead.

1 Introduction

Despite the impact of global insecurity, record debts and high oil prices, people's appetite for travel is unabated, with passenger numbers rising. In many ways, the difficulties faced by the sector since September 11 have served to demonstrate the agility and creativity of the travel industry. Each new challenge has prompted the arrival of a new service, an unexpected business model, a new way of working or a technological innovation. It is widely recognised that the best way for airlines to navigate this turmoil is to stay focused on their customers, the traveller. And as travel becomes cheaper, and purchasing power grows, the word *traveller* represents an ever growing variety of people, cultures, needs, expectations, aspirations and lifestyles.

Future Traveller Tribes 2020 is an attempt to understand how demographic, geographic, and political trends might shape travellers of the future. People from across the industry, as well as forecasting and technology experts, were used to identify existing and future groups – or *tribes* – of traveller. After a process of elimination, four tribes were identified. This was not because they were the only tribes expected to emerge, but instead because they were seen as having the biggest potential impact on how airline will deliver services in the future. We are already seeing these four groups emerging today.

Following an overview of the trends which have conspired to bring about the environment for these groups to emerge, the four traveller tribes will be described.

Importantly, alongside the description of these new customer segments will be an indication of the types of technology that airlines could employ to better serve their needs and build loyalty across these groups. Today's technology is evolving at such a fast rate that it is already set to have a huge impact on the future of the travel experience and the service that travel providers can offer. Communication technologies, distribution systems, sensing and identification technologies continue to advance. Harnessed properly, they will enable travel providers to deliver a more efficient, seamless and engaging experience at all stages of customers' journeys, at booking, check-in, in-flight and baggage collection. In some cases, the future is already here. Much of the technology in the report already exists, although it is not necessarily used as described.

2 Methodology

The methodology consisted of two elements.

Firstly, Henley Centre HeadlightVision carried out initial desk research to generate insights about air traveller tribes of the future. Sources included:

- Henley Centre HeadlightVision's proprietary global knowledge including:
    - *The Global Energies*, global trends in consumer and brand behaviour
    - *HenleyWorld*, a survey of consumer attitudes across 14 countries representing over 70% of global GDP.

- A variety of external sources including the World Tourism Organisation, United Nations, Organisation for Economic Cooperation and Development, World Bank, Official Airline Guide and the Association of National Tourist Office Representatives.

Secondly, workshops and interviews with Amadeus, airline representatives and other travel technology industry experts were conducted to validate the tribes and to generate insights into the technology implications and suggestions for each tribe.

3 Trends Impacting the Development of Future Segments

There are three key areas to focus on to understand the types of travellers which will emerge in the medium term; these are macro-contextual trends shaping the global environment, consumer trends shaping behaviour and technology trends.

3.1 Contextual Trends

What is striking is that the growth in global travel is expected to continue over the next decade, driven by a range of social, cultural, political and economic forces.

The burgeoning global population is set to increase to almost 8 billion people by 2025. Combined with the *growth in global migration* we will see a dramatically increased desire for international travel with many migrants remaining strongly connected to their country of origin with many reasons to return, such as visiting family and friends.

The globalisation of business is set to continue, if not hasten its pace. The rapid growth in world trade has led to increasing cross border traffic in goods, services and capital, with a global economy that is set to be 80% larger in 2020 than it was in 2000. This has resulted in a growing need for international travel to cement and manage international agreements and alliances.

Affluence is continuing to increase which has fuelled the growth in tourism. As people's material needs are fulfilled, their focus turns to enjoying experiences. Average income has grown substantially, particularly in the emerging BRIC (Brazil, Russia, India, and China) economies, and this is set to have a substantial impact on global travel in the future.

The globalisation of travel and tourism has ensured that the sector has emerged as one of the world's largest and most vital industries. The World Tourism Organisation predicts that international tourism arrivals alone will number over 1.56 billion by 2020, with average annual growth rates in Asia and the Middle East forecasted to be over 6% between 1995 and 2020.

The impact of Low Cost Carriers (LCCs) on consumer behaviour has been profound. LCCs have redrawn the air travel landscape, opening travel to groups that could not previously afford it. By using regional airports, LCCs are also now competing with traditional rail networks in many markets, but they are also set to extend into longer-haul journeys with a more extensive service, rather than exclusively focusing on their traditional short-haul service. Such advancements have enabled businesses to significantly cut the costs of international travel and have facilitated more face-to-face meetings.

New aircraft developments will continue to transform the customer's journey. For example, larger aircraft such as the recently-launched Airbus A380 will allow greater passenger comfort, while small aircraft will give greater flexibility on routes. In addition, more efficient technologies will enable airlines to fly cleaner, quieter and longer journeys.

The evolution of technology may reduce travel costs even further. The requirements of travellers and consequent complexity of information management will grow exponentially. To combat this, however, the development of efficient information systems architecture, the use of commodity hardware, open source software and the public internet as a communication network will enable travel providers to afford more streamlined and efficient information management technologies. As a consequence, they will be able to offer better prices and more comfortable and secure journeys to travellers.

3.2 Consumer Trends

Future travellers will be more sophisticated, knowledgeable and demanding. Affluence, education and choice will raise expectations.

The ageing population is a global phenomenon. By 2020, the world population aged 65 and above will have trebled from its present number, which is 700 million people. In many developed nations, falling birth rates and rising life expectancy means that older people account for an increasing proportion of the total population. At the same time, this ageing generation is healthier and fitter than ever before, and – at least for the moment – wealthier. This means that there is an increasing demand for travel and exploration from this group of well-off, older passengers.

We are seeing *changes in spending patterns* on two fronts, what is referred to as bargain-hunting and trading up. There has been a rise in people who are willing to 'trade down' in their travel purchases, looking for the cheaper option and even sacrificing quality or certain benefits for a lower price. Internet technology is the driver of this trend as it gives access to increased offer and price transparency. At the same time, increasing affluence and the media's preoccupation with wealth, fame and celebrity means that aspirations and expectations are rising. This has led to many people 'trading up' in certain purchases such as clothes and personal goods. In addition, what has become known as the 'democratisation of luxury' has led many consumers to expect premium service at affordable prices.

Everyone, it appears, is in a *search for health and wellbeing*. Changing global wellbeing will impact motivations for travel and traveller's needs. Leisure time has become a key opportunity to enhance wellbeing and find sense and escape in increasingly complex and hectic lives. A likely growth in medical tourism has also been predicted, driven by consumer's desires to avoid long waiting lists or to get access to procedures that are, relative to one's home country, free, cheaper, legal or which require advanced care that is not available in one's own country.

The *desire for personalisation* is a concept which is currently capturing the imagination of politicians, governments and organisations alike. The rise of empowered and affluent consumers in the West, and the rise of individualism, has created a culture of huge consumer choice. The traditional 'one size fits all' mentality is increasingly being challenged, as a growing number of consumers in developed markets demand solutions tailored to fit their needs rather than a mass market solution.

And, *sustainable tourism* is set to increase. More and more travellers are seeking responsible tourism, aimed at supporting the sustained development of local cultures and protection of the environment. Tourism that fosters benefits for the local people and their environment, while at the same time providing a fulfilling and exciting voyage, is increasingly in demand.

One concept that is emerging is what is known as '*modal lives*'. Multiplying demands and opportunities mean that modern consumers lead modal lives. The same person will occupy a number of roles within their everyday lives, and with these roles come different needs and demands. With regards to travel, those used to travelling business class on company missions may opt to travel economy class for leisure.

As levels of affluence rise, consumers' material needs are increasingly being fulfilled, and with these needs satisfied, consumers' interests have turned elsewhere. Increasingly they value experiences over possessions, both for enjoyment and to define who they are. Experiences such as travel hold greater kudos than material goods and have become the new way to express status. This is leading to what we call the *experience economy.*

3.3 Technology Trends

In an era when travellers want more control, comfort, security and personalisation, there is a real opportunity for technology to deliver something else – an engaging, human-centric interface with customers. Technology is now available that can identify and intuitively respond to people's individual needs, and this ***humanisation of technology*** will underpin the ability of travel providers to deliver exceptional customer experience in the future.

With this human-centric focus in mind, it is predicted that there will be four main areas of focus for technology development:

Digital personal identities (detailed customer information held digitally and therefore easily and quickly accessible), which will enable a far more personalised service. This can, and may, include;

- Tailored services and loyalty programmes that allow detailed passenger information to be stored and then incorporated into booking solutions
- Personalised destination information, to enable travel providers to ensure that travellers receive personalised health information, details of leisure pursuits and other information that will enhance their journey
- Portable access to digital identities. Easy customer recognition via devices that give access to personal details, containing information such as biometric details and travel itinerary, visa status and a Passenger Name Record (PNR).

Integrated information systems that combine information from a variety of sources;

- Generalised online and offline merchandising, boosted by new technologies which offer more choice and personalisation depending on travellers' profiles
- 'Natural shopping' for travel, where travel will be bought and sold as a commodity in regular retail stores
- Digital Concierge. Through the integration of advanced information systems, travellers will be offered a single point of contact that will enable them to deal with any travel-related enquiry
- Integrated consumer-facing sales systems which will expand the ability of travel companies to cross-sell travel and ancillary services to passengers including local transport connections.

Real-time, geo-relevant information delivered to individuals based on need and location;

- Mobile data service technology. Using mobile telephony, real-time journey management means travellers will be proactively informed of information relating to their journey
- Flexible journey management that will make use of technology, enabling significantly improved travel disruption management that will allow journeys to be seamlessly re-scheduled
- Tracking technologies for people and baggage. GPS and RFID technologies embedded in mobile devices or PNRs will enable the tracking of travellers and their baggage through their journey.

Increased customer interaction through social computing. A rising level of visual information technologies will be seen as they become cheaper and more sophisticated;

- 24/7 technology connections which will enable access to internet whilst on-flight or at the airport through mobile devices and/or wi-fi throughout the journey
- Advanced self-service kiosks that will be linked to digital identity, allowing check-in, itinerary and airport navigation prompts. Cognitive computing will enable customers' emotions to be interpreted and a relevant response delivered
- Wearable technologies including chip implants and RFID in clothing that will help identify the location of security targets and may also help track individuals through their journey, such as travellers who may need assistance
- Social computing such as increased peer communication online and affiliate marketing that will lead to consumers purchasing travel in different ways. Many of these technologies currently exist or are already under development.

However, they are expected to emerge as mainstream and ubiquitous in the future, although technology adoption will of course vary by region and carrier segment. Developments in each of these four areas of technology will have a profound effect upon the customer journey at all four key stages: booking, check-in, in-flight and baggage collection.

4 The Four Traveller Tribes

In the context of these three trends areas – contextual, consumer, technology – the following four tribes were considered the ones most worthy of further investigation.

Firstly, there is the ***Active Seniors***. These are people aged between 50 and 75 in 2020. Many of them will be enjoying greater wealth than previous generations and higher physical health, enabling them to take holidays and short breaks to relax and enjoy retirement.

Secondly, ***Global Clans*** reflect the growth in global migration and the resulting increase in the number of individuals and groups travelling internationally to visit friends and family for holidays, or to reconnect.

Thirdly, ***Cosmopolitan Commuters*** are the group of individuals who will be living and working in different regions, taking advantage of falling travel costs and flexible work styles to improve their quality of life. They will often live in one country or city and work in another and use air travel as the means to commute.

Fourthly, ***Global Executives*** represent the most affluent of the four traveller tribes, and comprise senior executives of companies with international interests travelling in premium class, increasingly by air-taxi or private jet.

Below is a more detailed outline of the four tribes – or customer segments.

4.1 Active Seniors

The first tribe – *Active Seniors* – has come about through a combination of factors. Advances in diet and medicine now mean that men and women are living longer, are healthier and are more active than previous generations. Many, importantly, will also be wealthier. Released from the constraints of work, affluent retirees and part-time workers will make the most of the wealth they have accrued and the freedom of retirement by taking holidays.

Defining characteristics of the *Active Senior*

- They will be youthful and adventurous seniors
- They will hail from affluent regions with ageing populations including most developed countries
- They will be seasoned and vocal consumers of products and services, who will have travelled extensively in their younger days
- They will likely seek holidays with a specific focus
- They are likely to look at shared interest and singles holidays that cater to this specific group, with many of them having experienced relation ship breakdown
- They will often research their journeys extensively, often doing so online, and planning well in advance to take advantage of good deals
- By 2020, there will be many more senior travellers from the emerging BRIC markets as their middle classes grow more affluent.

What are their key needs?

This can be summed up as *price, comfort and health.*

With finite funds available, *Active Seniors* are likely to be price sensitive but they will wish to balance these needs against a desire for comfort. Although some will travel economy class, a few may upgrade for special holidays or for longer journeys.

Mobility issues increase making longer journeys less comfortable, from distances travelled on foot at the airport, to time spent immobile on a plane. The acuity of senses declines with age, making reading and navigation harder, impacting travellers from point of research to airport navigation. Many will be managing long-term ailments and will have a sophisticated understanding of their personal health needs.

What is the evidence that *Active Seniors* will emerge?

By 2020 the world population of elderly people is expected to have trebled, with an estimated 700 million people aged 65+ years (more than 20% of the populations in France, Germany and Japan will be over 65).

Forecasts predict that GDP in Eastern Europe will increase by nearly two and a half times between 2005 and 2015 to US$ 4.49bn, whilst GDP is set to more than double in the Asia Pacific region over the same period.

A survey carried out by NOP, the UK research agency, for Saga, the over-50s service provider, in 2004 found that two-thirds of respondents would travel the world and 1 in 10 would consider a trip to Antarctica.

What are the technology implications?

There are a number of possibilities. Some of the most obvious implications are detailed below.

1) Customised travel itinerary at booking stage e.g. personalised itineraries, allowing travellers to have more control over their travel experience. This would enable them to make specific booking requests regarding seat position, or to pre-order reading material or in-flight entertainment that helps them experience their travel destination in advance of arriving.

2) Dynamic Visualisation such as virtual reality and other high quality visual imagery to allow travellers to sample the travel experience in advance of booking the tickets.

3) RFID baggage tracking at check-in stage to facilitate bag tracking at the airport itself and throughout the journey, reducing the need for less physically able travellers to struggle with their luggage. This could be combined with an onward delivery service to their accommodation.

4) Health sensing technologies to be worn during the flight to monitor certain aspects of health or wellbeing such as heart rates or anxiety levels. In response, elements of the internal environment are altered to reduce stress points or to respond to specific health needs.

5) Personalised healthcare services that can identify passengers with specific health needs before they board the aircraft, and allow passengers' medical records or prescription details to be available to a designated doctor (who can speak their language) at their destination.

In the longer-term, there may be further developments available such as *digital memories management*, to enable them to recall and potentially repeat what they did and experienced in the past or potential *reconfiguration of in-flight seating arrangements* according to the mobility and space requirements of passengers, based on an assessment of passengers as they board the aircraft.

4.2 Global Clans

The second tribe – *Global Clans* – is emerging due to increasing global migration. As more people migrate in search of better prospects and lifestyles, more people than ever before will be travelling internationally, either alone or in family groups, to reconnect with family and friends.

Defining characteristics of the *Global Clans* member

- They will be most strongly associated with regions with large immigrant populations. The key immigrant destinations are currently Northern America, Europe and Oceania, with most travelling from Asia, Latin America/Caribbean and Africa
- They will be embedded within their host countries. As such, their attitudes and behaviours may change, with second generations showing hybrid lifestyle aspirations and cultural attitudes reflecting both the country of parental origin and the new homeland
- The links between families will remain strong as international communication via the internet and mobile technologies becomes cheaper and more accessible
- They will likely organise travel to coincide with key dates such as national holidays and festivals when families traditionally reunite
- They may also have to travel to respond to unpredictable family issues
- They will increasingly book online, often in large, coordinated group bookings.

What are their key needs?

This can be summed up as *prices and group travel.*

Price sensitivity is key to *Global Clans* as many will be less affluent, with their focus for travel being family reunion rather than a 'holiday experience'. Therefore many will be likely to compromise comfort for cost savings.

They are likely to travel in economy class and many may be willing to travel at non-peak times, including night time to gain cost savings.

As travel will largely be on key dates, bookings will be made in advance, except on the occasions when there is a necessity for immediate, unplanned travel due to unforeseen circumstances such as illness.

Global Clans will be more focused on groups than the other three tribes, perhaps wishing to book and travel together. Within these groups, different members will have different needs, such as children versus grandparents.

What is the evidence that *Global Clans* will emerge?

There are currently 191 million people living outside the country of their birth.

Over the last 25 years the number of international migrants has doubled. The UN expects that international migration will grow to 250 million by 2050. The largest Japanese immigrant population can be found in Brazil.

The Pew Internet and American Life project (2006) found that the internet expands people's social networks and even encourages people to talk by phone or meet others in person.

What are the technology implications?

1) Web-based group forums at the booking stage whereby travel providers could offer online forums to facilitate group communication about booking dates and travel plans.

2) Cross-regional booking to allow family members in one region to book travel on behalf of family or friends in another country.

3) Accessible and secure payment and booking, to include innovative payment solutions that would eliminate some of the security risks associated with booking by credit card or using cash in certain regions, e.g. by the use of electronic travel vouchers.

4) Booking booths and new distribution channels in convenient locations such as supermarkets or music stores, to allow customers to book journeys more easily, with the journey details to be emailed or sent to mobile communications devices.

5) Additional cargo/luggage services to allow those with special luggage requirements to book their luggage on mini-cargo planes in advance.

6) Airport navigation; the installation of touch screen kiosks that operate in a number of languages providing location information and guidance through the airport.

4.3 Cosmopolitan Commuters

The third tribe – *Cosmopolitan Commuters* – is a response to changes in both social and economic conditions. Motivated by the promise of an improved quality of life, a growing number of people will be taking advantage of flexible working arrangements and the relatively low cost of travel to live and work in different regions, for example living in Nice, France and working in London, UK.

Defining characteristics of the *Cosmopolitan Commuter*

- They will live in areas which they consider to provide a higher quality of life and where their salary will go further, or where the lifestyle better fits their priorities
- They will commute short-haul distances to work on a weekly or fortnightly basis, depending on their work commitments
- They will most likely be executives, freelance consultants or employees with flexible work contracts predominantly in their 20s, 30s and 40s, who will be able to work remotely due to the cheap, fast communications technologies available today
- They will often be driven by a desire to progress in the workplace by taking advantage of the rising numbers of short-term international placements offered by their employers, in order to increase their experience and market knowledge
- They will travel at peak times during the week, and their travel is likely to be frequent and block booked in advance for cost saving, although occasionally due to unexpected business meetings, bookings will be last minute and less price-sensitive.

What are their key needs?

This can be summed up as *time efficiency and flexibility*.

Time-pressed *Cosmopolitan Commuters* will not wish to waste unnecessary time travelling. Hence, the convenience of travel hub location, the ease of booking and the speed of check-in and security will influence their choice of travel providers significantly.

Cosmopolitan Commuters will need to be able to work on the go. Therefore access to technology, both during travel and at departure lounges, will be vital for this group.

As business meetings will change, booking and ticket flexibility will be an important need for travel providers to consider, particularly for this frequent travelling group.

Down-time will be an important need for which travel providers must cater, as commuters return stressed from business meetings or wanting to wind down for the weekend.

What is the evidence that *Cosmopolitan Commuters* will emerge?

The growth in international commuting will be facilitated by a decline in fixed hour jobs: a survey for the European Foundation for the Improvement of Living and Working Conditions in 2006 found that flexible working time arrangements exist in about one in two (48%) establishments with 10 or more employees in Europe.

63% of interviewees in a European survey, reported at www.personnelzone.com, predict an increase in international commuting where employees work in one overseas location and commute to another.

Research from the Recruitment Confidence Index found that 42% of respondents say they have increased their international recruitment over the past five years, nearly one in five (18%) say the number of staff leaving to work abroad has also increased over the past five years, and 20% of firms say it is important or very important that their senior managers spend time overseas.

What are the technology implications?

1) Real time ticket price information at the booking stage to be supplied via reservations technologies built into the calendar functions of handheld devices highlighting the most cost-effective, convenient travel arrangements for different times and dates across a variety of forms of transport including rail and air travel. Real-time, online comparison of ticket prices for the same route at the same time would allow passengers to select the best deal.

2) Bundling services and up-selling to enable the purchase of customised packages, for example the purchase of frequent flight 'coupons' in bulk, or a menu based system allowing access to the business lounge – even if booked on an economy flight.

3) Web-based digital history (personal and travel records) to enable travellers to input their preferences, in addition to storing their travel records. In the future, this will allow their needs to be recognised at the booking stage, leading to the proactive provision of added services.

4) Environmental awareness. Airlines to offer passengers the opportunity to offset the carbon emissions of their flight, perhaps by paying extra on their ticket price.

5) Flexible journey management to allow those who arrive at the airport ahead of time for their scheduled flight to easily board an earlier plane following the 'forward flow' concept. This will streamline the travel experience for the time-sensitive traveller.

6) Optimised security to make it possible to save passengers' time by informing them in advance of how much time they should allocate to security checks or by giving a personal time slot for security clearance to speed up the process.

7) 24/7 connections incorporating internet, VoIP, mobile phones and integrated laptops to enable people to continue working during travel.

In the longer-term, the overall streamlined travel process may also incorporate *new generation self-service kiosks* and *combined 'chip-and-go' passport and multiple PNR devices.*

4.4 Global Executives

The final tribe – *Global Executives* – covers the high value business travellers seeking a premium experience which is expected to increase as globalisation drives up international trade. *Global Executives* will be senior executives who will either travel alone, with an entourage, or with a partner, perhaps combining a business and leisure trip.

Defining characteristics of the *Global Executive*

- They will typically originate from developed markets around the globe, but by 2020, growing numbers will be from the BRIC countries with rapidly expanding economies
- They may travel premium or business class for some trips, but they will increasingly favour the more exclusive, time-efficient, bespoke travel services allowed by private jet and air-taxis
- Their main purpose for travel will be international business meetings and events, although leisure may also occasionally be integrated into travel
- They will not be price-sensitive and they will not manage their own travel booking
- Their trips will be a combination of long-haul and short-haul depending on their business.

What are their key needs?

This can be summed up as *time efficiency* and *flexibility*, the same as the Cosmopolitan Commuter.

Global Executives will be used to premium, luxury experiences and will have high expectations, often expecting a highly personalised service. Time is precious to Global Executives who will demand a highly streamlined and time efficient journey at all stages of their trip.

Global Executives need to have access to a wide range of technology, both in-flight and at airport departure lounges.

As business meetings will change, ticket flexibility will be important depending on ever-changing business needs.

What is the evidence that *Global Executives* will emerge?

According to the latest Barclaycard Business Travel Survey nearly half (45%) of business travellers say they are or were travelling for business more in 2005/06 than they were in 2004/05. The main reason given for the increase in business travel was business expansion overseas (33%). The survey also reveals that an increase in global business travel will boost the use of premium services. By 2015, the number of business travellers using business class services will rise from 17% in 2005/06 to approximately 33% in 2015, driven by the demand for improved service and better working conditions.

The number of people taking seven or more long-haul flights per year has been predicted to increase from 6% in 1997 to 14% in 2015.

With long haul travel increasing, the number of flights taken per business traveller is predicted to increase by 12% from 7.6 flights per year in 2005/06 to 8.5 flights in 2015.

What are the technology implications?

1) Superior flexibility to provide last-minute multi-channel booking, re-booking and/or cancellations on the go.

2) Accessible web-based digital history (personal travel records) to allow the pre-emption and pre-identification of needs based on their travel history and passenger profile even if they do not fly frequently on the given carrier.

3) Personalised premium journey packages driven by the predicted growth in demand for private jets, tailored packages for *Global Executives* could be constructed in alliance with private jet operators.

4) Information en-route to be supplied by travel providers offering podcasts that can be downloaded en-route to travel hubs such as airports, providing up-to-date information on road construction, check-in status etc.

5) Fast-track lane, cross-function check-in and security counters with integrated information will save time and hassle at airports.

6) Personalised welcome facilitated by GPS technologies may advise staff at departure lounges of a *Global Executive's* imminent arrival.

7) In-flight networking would provide the perfect opportunity for business travellers to network with others. Information or electronic business cards could be circulated allowing passengers to identify potential networking opportunities on their flight.

In the long-term, there may be the potential for *advanced pricing models* that would enable jet operators to arrange the scheduling of flights to provide certain cost reductions if passengers are able to be more flexible with timings, or if *Global Executives* are willing to group-share flights to the same destination at certain times.

5 Conclusion

There were two major conclusions from the *Future Traveller Tribes 2020* report.

Firstly, the demand for travel services will continue to increase despite changing geo-political, economic and social conditions. The four tribes will be at the forefront of the changing type and level of services that airlines provide. And each one of them poses a unique set of challenges and opportunities that need to be considered. What is clear is that the traditional leisure/business segmentation will no longer satisfy the increasing expectations of travellers today or in the future.

Secondly, it is clear that technology has a major role to play in bridging the gap between travel providers and travellers. The need for the *humanisation* of travel technology is a major theme to emerge from this exploration of future travel expectations. Technology needs to be developed to support people with a wide variety of backgrounds and requirements when they are travelling, in a way which enables high degrees of personalisation and exceptional customer service.

Innovations in Destination Distribution Management

Walter Freyer, Manuel Molina

1 New Marketing Channels in Tourism

Tourism distribution systems have been changing dramatically over the last several years – this is reflected not only in traditional distribution channels such as tourism suppliers and tour operators but also in destinations. The business of the traditional travel agencies in particular is undergoing transformations: as a result of the newly modified terms of competition, most of the independent travel agencies had to reorganize and join travel agency chains and affiliations within a very short time span. In 1990, the sum of unaffiliated travel agencies in Germany was around 80% whereas in 2007 this number has dwindled to below 5%. Currently, the role and function of the classic stationary travel agency and the new online marketing channels are under discussion.

At the same time, many tourism suppliers are also looking for new distribution channels. As such, tourism destinations are increasingly running into competitive pressure and are thus being driven to introduce innovative marketing policies as well as redefine their marketing channels. Up until recently, destination marketing had been left in the hands of national and regional tourism organizations as well as incoming agencies and tour operators. However, the increasing competition between destinations inevitably requires adopting a professional and modern design for distribution and marketing channels. In fact, many destinations have already realised that the marketing distribution offered through traditional tour operators no longer fulfill their needs. Thus in order to reach out to and stimulate end consumers and retailers, these destinations are increasingly opting to become more active. Innovative travel agencies offer destinations entirely new channels through direct marketing. By taking advantage of this offer, destinations can successfully tap into the massive marketing power held by tourism distribution.

If destinations are ranked according to the classic tourism distribution structure, they will always have hybrid or multiple functions. Destinations are to be considered service providers through their natural and man-made supply (such as transport and accommodation companies at the destination). They also take on the role of travel agents by arranging accommodation and thirdly distinguish themselves as tour operators through their travel packages.

Several of these aspects will be dealt with in the subsequent text. The changes in the marketing channels, from single channel marketing to multi-channel marketing, will be identified and the most important reasons for these changes will be specified. Shown by the example of the travel affiliation TSS new innovative perspectives and its chances for destinations will be demonstrated.

2 From Single-Channel to Multi-Channel Marketing

For many years, marketing was focused on only one distribution channel. Most of the time, an optimal distribution channel was sought based on the principle of "one-producer/product-one-channel" (Ahlert 1996; Meffert 2000: 660 et seqq.).

The options for distribution policy included (simplified):

- direct or indirect marketing (degree of directness),
- internal or external sales channel (type of intermediary),
- single-tier or multi-tier marketing (number of vertical distribution tiers – "length" of the distribution chain),
- as well as intensive, selective or exclusive distribution,
- in addition up to nine channel flows will be considered as functions of the distribution system (Kotler/Bliemel 2001: 1076 et seq.)

Until quite recently, multi-channel distribution was actually the exception to distribution policy options. Multi-channel distribution was only considered an option when a single channel did not serve to exhaust the market potential, when the marketing costs could be reduced using a different channel or when a Special customer access had to be attained. In such cases, an isolated multi-channel approach based on variety seeking was implemented: the objective was to address various customer groups using various marketing channels.

In general, authors indicate that similar customer groups may be exposed to cannibalization, "channel war" and high coordination costs (Wirtz 2007: 21 et seqq.), see Table 1.

Initially, sales and marketing in the tourism sector were clearly structured. Until 2000, most tour operators and service providers (transport and accommodation companies) depended on traditional marketing channels provided by stationary travel agencies (Freyer 2007: 496ff).

In general, the alternative was direct marketing; however, this was rarely seen in the private sector of the tourism industry, usually only in specialized tour operators for example.

In contrast, tourism destinations predominantly tended to rely on direct marketing provided through regional and national tourism organizations. In addition, destinations also received marketing benefits through tour operators and were thus able to use the classic marketing channels provided by these organizations. On the rare occasion, this co-operation extended to the point where destinations participated in joint marketing activities. Unless destinations provided bookable package tours, the implementation of traditional travel agency marketing actions was the exception. In this phase, destinations could only slightly perceive the marketing potential presented.

Up until the end of the 90s, the original value added chain in the tourism economy flowed from service provider through tour operators and travel agencies to the end-consumers. CRS were primarily established between service providers and tour operators, and later found their niche between tour operators and travel agencies. Later, this structure transformed: CRS and Internet began appearing between all stages, at first as medial support and later as self-sufficient institutions.

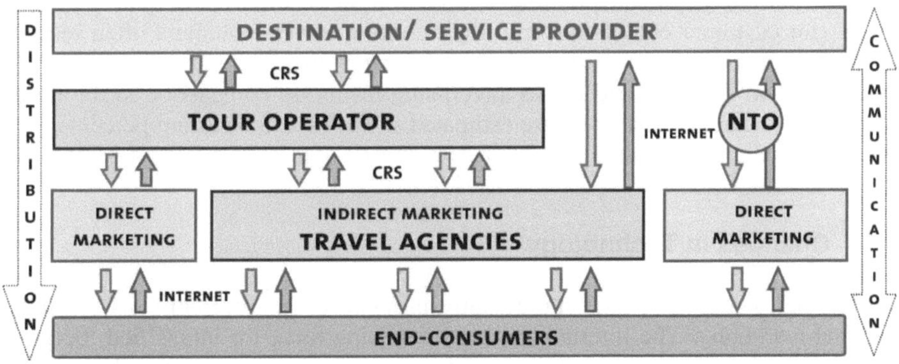

Fig. 1: Classic destination marketing

3 Reasons for the Change

Classic distribution channels began changing in the mid 90s. These changes were not solely limited to tourism but also encompassed other economic sectors. Three factors in particular, are often noted as causal arguments. In the tourism industry specifically these aspects manifest with slight modifications.

After the liberalisation of the restrictive agency contracts in Germany, many new opportunities arose for tour operators and travel agencies. Likewise, a new consumer behavior pattern emerged, significantly influencing the tourism distribution structure. In addition, a specific trend developed in the travel and tourism industry in which the initial distribution channels transformed to include information and communication channels. And, with the expansion of the Internet, new technical possibilities arose, hailed Online or Internet marketing.

3.1 Changes in Consumer Buying

First of all, purchasing habits of consumers in general (and of tourists in specific) changed on the demand side. Modern consumers can be characterized as multi-optional customers and consumers, who tend to perform very diverse buying behavior patterns; sometimes simple and cheap, sometimes expensive and exclusive. Such behavior is also mirrored in tourists and guests, who use several channels at the same time to complete their purchases ("channel hopping", Wirtz 2007: 281): stationary travel agency, direct contact with airlines or hotels as well as the Internet. Through these actions, guests verify which of the marketing channels can best fulfill their expectations. In general, it comes down to options, information, price, speed, reliability of the reservation etc.

Different channels are also used in tourism depending on the phase in which the purchase/reservation process is. Consumers often prefer using the Internet (online sources) during the information phase, stationary travel agencies (face to face) for reservations and payment, possibly written communication in the post purchase phase (for customer complaints) etc. Furthermore, travel customers often request catalogues and visit tourism trade shows, as long as they continue to be perceived as distribution channels and not as advertising media. In comparison to the USA, where multi-channel customers are estimated at 35–70%, the current percentage in the German market amounts to approx. 10% (Wirtz 2007:325, Fritz 2004:119).

3.2 Changes in Technology

Technological changes have significantly influenced the developments in multi-channel marketing. The Internet has been a driving force for intensified focus on multi-channel marketing. In addition, the influence of continuing development in CRM systems and their Internet interfaces is an equally important factor not to be forgotten.

Online marketing and e-commerce promote changes within the tourism value chain by reorganizing the tasks associated with the distribution channels and tools. Originally, the electronic approach to distribution was viewed as a special form of a distribution instrument, rather as "technical" support for face to face, written or telephonic contact to customers.

Initially, the Internet served primarily as a channel of information during the purchase decision, whereas the actual purchase often occurred in a traditional travel agency. In the meantime, the Internet behavior of the multi-optional consumer has become much more complex. The practice of using "as many (marketing channels) as possible" is no longer viable, a "targeted and strategic" approach has become standard procedure.

Today, e-commerce is viewed more and more as an independent channel, which competes with the classic tourism distribution channel, travel agencies, through in-house electronic companies or departments. As such, e-commerce has had a heightened institutional impact in tourism sales and distribution.

During this process, airlines were able to secure a cutting-edge position. Since airlines offer a product that does not require substantial consultation, its online distribution was quick to penetrate the market. As a result, the large travel and tourism portals emerged – partially still under the patronage of the airlines.

Agencies which depend solely on the online approach, use the Internet platform as single-channel marketing. However, many travel agencies use it to supplement their multi-channel marketing: On one hand they can provide service 24/7 and on the other they can gain new customers, who would not have made purchases at a stationary agency.

3.3 From Channels of Distribution to Channels of Communication and Contact

When regarding the traditional distribution of goods the focus is on, logistic problems: How does the product come from the producer to the consumer? For the most part, one-way or pipeline distribution prevails. In contrast, the predominant services provided in tourism, sales and communication, are tightly intertwined; therefore, two-way communication or dialogues dominate the scenario.

Theoretically speaking, this is the *uno-actu* principle of production and consumption, whereby the external factor (the guest) becomes an integral part of the service rendered, as well as face to face contact is given. Instead of one-way distribution, communication channels or – more generally speaking – marketing channels are to be found. In addition, new media are also progressively offering feedback – and thus two-way communication instead of the traditional one-way or pipeline exchanges found in the classic distribution system for material goods.

A transformation or a fusion of traditional distribution channels and media is increasingly emerging in multi-channel marketing. For example, every sales or contact approach occurs via a particular (communication) medium, i.e. face to face, written, electronic exchanges. Some of these former communication media convert into "new" distribution channels and institutions, i.e. call centers (contact via telephone), direct mailings (letters: written communication), Internet (electronic contact).

The most important communication (medium) and distribution (institution) channels in tourism include:

- **channels via media** (communication channels): face to face, telephone/fax, PC/online/Internet, PDA/cell phone, catalogue/brochures, radio/TV advertisements, printed media, direct mailings.
- **channels via institutions** (distribution channels): stationary travel agency/agent, e-commerce, m-commerce, sales representatives (door-to-door or traveling sales), market events (trade shows, events).

The classic one-way or pipeline thinking in distribution ("special distribution channels for specific customers") is undergoing profound changes. More and more, the distribution channels are developing into two-way (interactive) channels of communication (including interaction functions) – thus the most suitable term seems to be *marketing channels*.

Within this context, the Internet, with its multi-media facet, has played a significant role in the distribution of tourism products. Potential travelers gather the best possible information about a specific product through visual hotel tours, webcams which provide images from the destination, travel reports etc.

Information and reservation options are continually merging. As such, destinations in particular are put under pressure to act. Instead of providing a simple information platform, destinations now yearn to promote reservations. The travel agency affiliation TSS for example offers a possible one-stop solution.

4 Multi-Channel Marketing in Tourism

4.1 Multi-Channel Marketing Is More than Online Marketing

Multi-channel marketing in tourism is much more than just online marketing, even though e-commerce is one of or even the most important component. The terms online, Internet as well as e-commerce and e-business will be treated in this article as synonyms, even though they may have slightly different meaning (Fritz 2004, Eggert 2006).

The key idea of modern multi-channel marketing in tourism is to approach customers over various channels. It is often stated that marketing and distribution are to be structured so that they reach the customers wherever they should be. As such, different channels should be created for different customers.

The Internet complements and enriches stationary marketing rather than replacing it. At the outset of Internet marketing, travel agencies saw this channel as direct competition. Now, they take advantage of its full potential.

In tourism, different marketing and communication channels are consulted during the different phases of the travel decision-making process, known as *customer touch points* (Freyer 2007: 508ff).

- **Attention phase**: Advertisement (in classic media, shop windows, events or online), an actual desire for a trip arises
- **Information phase**: Consulting a call center, travel agents and/or dialogues (online, TV)
- **Reservation phase**: Making reservations in a stationary travel agency
- **Transaction/Pre-trip phase**: Payment transaction, ticketing in a travel agency, sometimes even online
- **Travel phase** ("en route"): Support through service providers and the incoming agency (possible calls regarding customer complaints from the destination), additional reservations (excursions), additional bookings
- **After sale/trip phase**: Customer complaints, postproduction/contacting the travel agency, writing the tour operator

4.2 Integrative or Comprehensive Multi-Channel Marketing

Implementing various channels in multi-channel marketing can bring about promising results; however, to achieve this success, various channels have to be properly coordinated and integrated as distribution and communication channels. In addition, they must also cater to complex consumer expectations. In other words, distribution and communication have to merge – to integrate. Such integrative or comprehensive multi-channel marketing does not try to unite numerous independent distribution groups. It rather endeavors to offer as many "one-stop" approaches as possible. Many refer to this approach as *one face to the customer*. In such a system, the marketing channels have to be coordinated in order to evoke basic confidence and recognition. The goal is to lure in the elusive "hopper". Such is the case in particular in the marketing concepts of sales organizations like TSS. Multi-channel marketing has transformed from either-or-marketing to as-well-as-offers.

Important channels within multi-channel marketing include:

- the traditional stationary travel agency – provides security and ensures customer proximity as well as face to face contact
- the online travel agency (with Internet brands)
- the call center – to advise and support customers

as well as:

- catalogues/brochures (as informational material, product presentation etc.) trade shows and events
- traveling sales / door to door sales – approaching the customer
- face to face: attending market events (street festivals etc.)
- mouth to mouth: as part of Web 2.0

By these channels, the whole range of marketing (ideally with a one-stop structure) has been defined and the extent of marketing power is available!

4.3 Advantages of Multi-Channel vs. Single-Channel Distribution

Multi-channel marketing provides many advantages in comparison to single-channel marketing and has created new mechanisms for customer retention. In fact, the development of new marketing channels often translates to new customers and thereby additional business volume. Through multiplier effect cross selling can be created. Linking these mechanisms together can therefore result in improved cost-effectiveness in sales and communication. Other positive outcomes include establishing a balanced, comprehensive brand and image development as well as controlling management. Furthermore, multi-channel marketing ensures greater market penetration, since contact to customers transpires through various communication media and can therefore tap effectively into market potential.

Table 1: Opportunities and risks of multi-channel strategies

Opportunities	Threats
• larger market penetration • comprehensive customer support (incl. CRM) • cost-effectiveness: optimizes the distribution costs, balance of risks • cross-selling possible • competitive advantage with positive integrative image and brand conception	• channel conflicts (channel wars, cannibalization effects) • customer confusion, partially due to a misfit in regards to the CRM • high coordination costs, highly complex • high set up costs in reorganization • loss of control, sub-optimization

On the other hand, the aforementioned risks, including channel conflict, high coordination expenses (and therefore higher costs and complexity) as well as a possible sub-optimization associated with these costs and the loss of control still exist. In regard to customer relationship management (CRM), an undesired "misfit" could arise if the integrated and comprehensive channel relationship is not given.

5 Destination Marketing – New Challenges and Opportunities

5.1 The Three Stages in Destination Marketing

Modern, professional and active destination marketing has just begun to develop and is relatively new. Quite often a trend of passivity still prevails with the motto "wait until the guest arrives". In this first stage, the function of destination marketing encompasses providing information in response to customer inquiries. The innovation in destination marketing focuses simply on attending trade shows: the main goal consisted of distributing brochures to promote the destination. Since most of these organizations are public, competition and thus the orientation towards the market/customer were fairly limited.

A transition towards active destination management first began to arise at the end of 2000. In this second stage, many destinations discovered the advantages of working in cooperation with tour operators etc. and began strengthening these relations with joint marketing activities (i.e. events and road shows). This was possible because many destinations opened offices abroad. Now they were closer to the source market but in most cases still didn't know about the wants and needs of their customers.

In the third stage, many destinations opted for reorganization. Since their products are to a certain extent interchangeable and since the pressure of competition is continually rising, destinations have to take an active stance in marketing. At this point, destinations recognize their responsibility to provide information and products and begin to work actively to capture the attention of their source markets. Initially, many destinations discovered the advantages of the Internet and gradually expanded their Internet presence on their own initiative. Professional marketing strategies and concepts are essential for modern, and of course, market-driven destination marketing or even marketing policy. For the first time, new marketing partners such as call centers began to appear along side the traditional marketing form, NTOs and tour operators.

5.2 Innovative Destination Marketing Using the TSS Multi-Channel Model

Travel agency affiliations such as TSS, registered in Dresden with approx. 2,300 independent travel agency partners, have recently begun to offer new possibilities and solutions for destination management based on sales and marketing in specific source markets. Marketing organizations thus provide destinations with platforms and channels based on a multi-channel system.

The goal of modern marketing is to always be where the customer is. With their multi-channel strategy, TSS has developed an approach to redirect sales, which are increasingly being generated outside travel agencies, back to travel agencies. The reason for developing a multi-channel strategy in the TSS Group was to provide travel agencies with an innovative approach and competitive concepts. This strategy provided agencies with a foundation to react to the structural transition in the tourism industry and to successfully position themselves in the market. Since its foundation, it has grown into a marketing structure with versatile platforms, from which even destinations can gain market value. The model consists of modular marketing channels, offering multiple combinations for optimal sales and marketing approaches.

The TSS multi-channel model consists of the following marketing channels:

1. *Stationary distribution* as the core competence (approx. 2,300 affiliated TSS travel agencies sell traditional tour operator services, dynamic packaging etc)
2. *Online sales* over the brand Onlineweg.de (Internet presence in travel agencies. In addition, travel agencies can include a subdirectory in the main portal and thereby gain additional profit through commission rates and marketing activities organized by the main portal).
3. *Call Center System* with B2B and B2C sales (TSS Service Center)
4. *Travel production / sales* through the operator brand "travelers friend" (operators with attractive commissions)

How can a destination profit from multi-channel marketing provided through travel agency affiliations?

The main advantages to this approach are access to a long standing and established sales and marketing structure with various channels and, as a result, direct representation at the point of sale.

Consequently, through the contact with the one partner association (the co-op, which acts as a representative), destinations are able to reach a multitude of travel agencies, who have acquired knowledge about travel destinations (i.e. in training programs, with print media etc.) and then distribute this information to end consumers.

Aside from distributing information, destinations can also directly market and distribute their products. Other instruments for distribution include call centers or the organizer's brands as well as diverse Internet platforms.

Destinations are thus able to successfully achieve a viable link to the source market, generate direct demand or influence consultation and purchasing decisions.

The complex duties and responsibilities of a destination as producer, marketing and sales agent create very high standards for the cooperation with various source markets. As such, market-specific know-how and concepts are essential. For this purpose, marketing organizations such as TSS have already responded to the market demand and have begun providing relevant services.

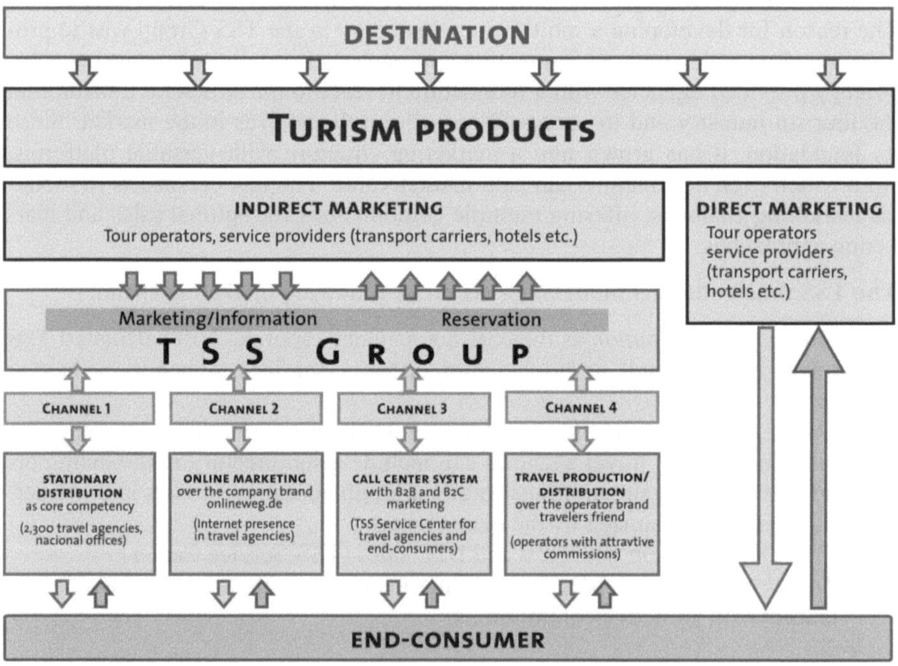

Fig. 2: Marketing channels within the TSS Group multi-channel strategy

References

Ahlert D (1996) Distributionspolitik: das Management des Absatzkanals. 3rd edn. Urban & Fischer, Stuttgart
Egger R (2005) Grundlagen des eTourism; Informations- und Kommunikations-technologien im Tourismus. Shaker, Aachen
Fritz W (2005) Internet-Marketing und Electronic Commerce. 3rd edn. Gabler, Wiesbaden
Freyer W (2007) Tourismus-Marketing, 5th edn., Oldenbourg München/Wien
Homburg C, Schäfer H., Schneider J (2006) Sales Excellence – Vertriebsmanagement mit System. 4. Aufl., Gabler, Wiesbaden
Kotler P, Bliemel F (2001) Marketing-Management, 10th edn. Schäffer-Poeschel, Stuttgart
Meffert H (2000) Marketing 9th edn. Gabler, Wiesbaden
Schögel M (2001) Multichannel Marketing – Erfolgreich in mehreren Vertriebswegen. Werd Verlag, Zürich
Wirtz B (Ed.) (2007) Handbuch Multi-Channel-Marketing. Gabler, Wiesbaden

Travel Technology and Business Travel Management

European Online Travel Overview – An Abstract of PhoCusWright Online Travel Overview Third Edition

Michaela Papenhoff, Klaus Fischer, Roland Conrady

1 Introduction

The European market is the largest regional travel market in the world backed by some of the strongest travel and Internet infrastructure elements in the world. The European online leisure/unmanaged business travel market as a whole averaged more than 50% annual growth from 2003-2006. But the gains have slowed, and will continue to cool through 2008, when the online channel is expected to increase by 22% compared to 37% in 2006 (see Fig. 1).

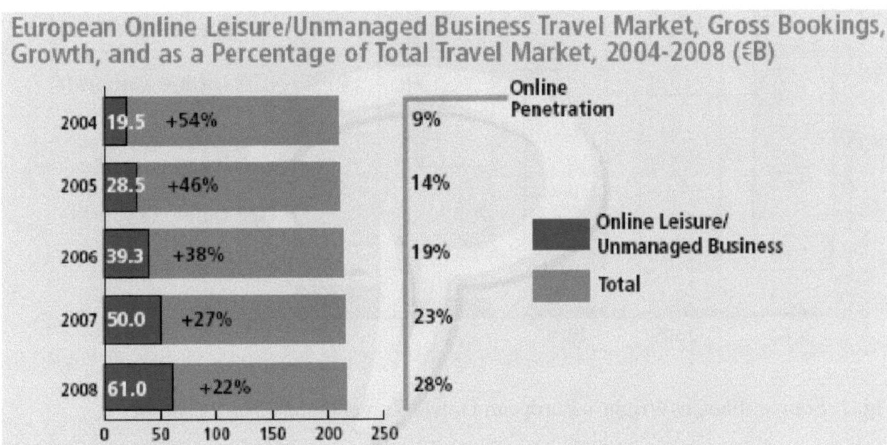

Fig. 1: Source: PhoCusWright's European Online Travel Online Third Edition

Despite the tempered growth, the channel remains robust: the online leisure/unmanaged business travel market in Europe approached €40 billion in 2006. Corporate Internet bookings showed improvement as that market matured, particularly in the U.K. and Germany, but consumer-facing Web sites are driving growth in the channel and will continue to do so for the foreseeable future. Leisure/unmanaged business travel gross bookings have dominated online sales. In 2008, the leisure/unmanaged business segment, currently nearly seven times the size of the corporate Web segment, will still be more than triple the size, €61 billion versus €17.2 billion (see Fig. 2).

In the worldwide travel market, the European travel market has not claimed the same share of online travel as it has of the overall travel market, an indication of the depth of potential that exists in the channel in Europe. As that potential is exploited, a balance will be achieved, and by 2008, Europe's share of the online travel gross bookings (36%) of the three major economic regions of the world – the U.S., Asia Pacific region (APAC), and Europe – will match its share of total gross bookings in the regions (see Fig. 3). The change is notable: in 2005, Europe contributed 39% of all travel gross bookings in the three regions (which comprise close to 80% of all travel revenue in the world), but only 27% of online travel gross bookings for these three regions were European.

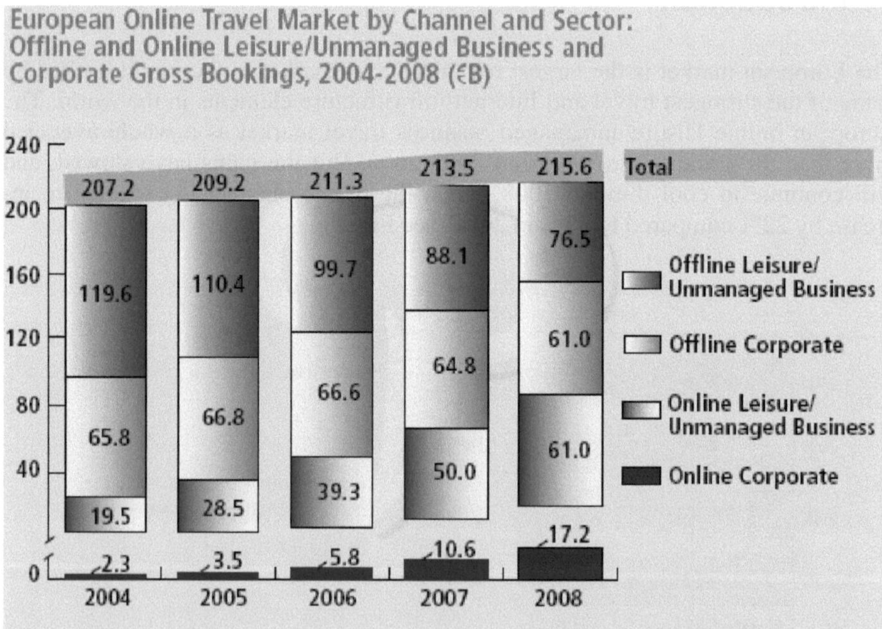

Fig. 2: Source: PhoCusWright's European Online Travel Online Third Edition

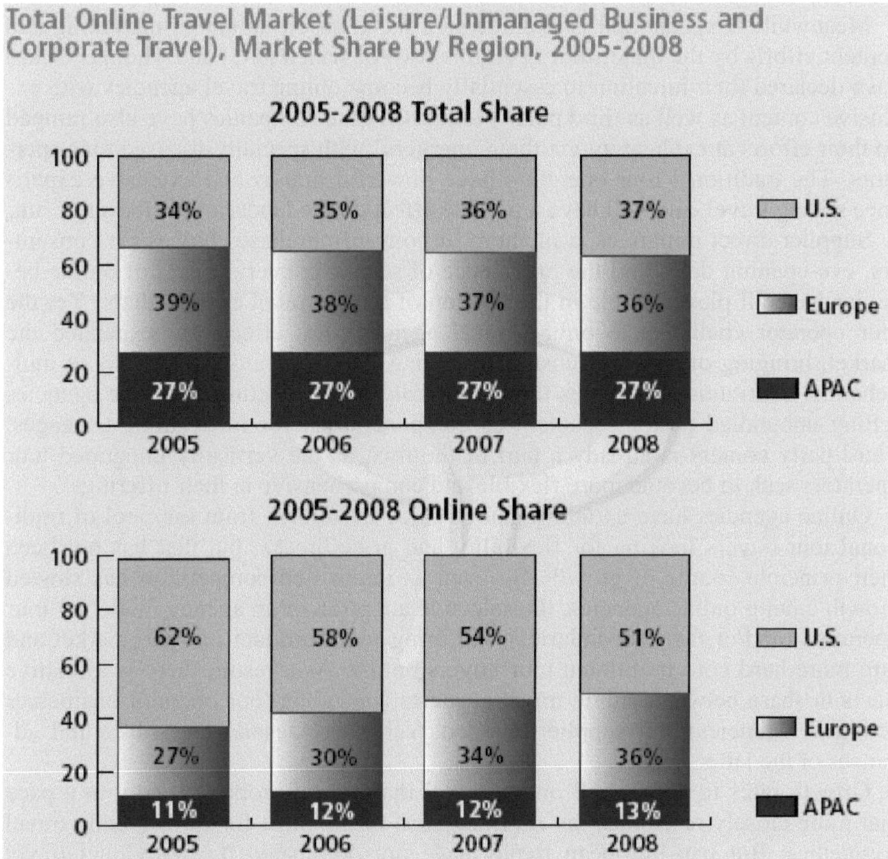

Fig. 3: Source: PhoCusWright's European Online Travel Online Third Edition

2 Online Agencies Face New Realities, Intense Competition

Being everything to everyone is no easy task, however, as the online agency giants can attest. Expedia, lastminute.com and Travelport's ebookers have all seen their share of troubles, deriving from a variety of sources but reflective of the complexity of serving a diverse customer base. Travelocity's 2005 purchase of lastminute.com may have masked some of the slowdown in the online agency channel, but there was no hiding ebookers' woes earlier that year. These businesses were impacted by the difficulties of managing offline businesses while integrating online offerings.

Meanwhile, Expedia and lastminute.com faced an onslaught of marketing and content efforts by the major tour operators, two of which (TUI and Thomas Cook) have declared their intention to essentially become online travel agencies with exclusive content as well as third party products. These companies have also ramped up their efforts at exclusivity via their "mergers" with specialty-focused tour operators. The traditional tour operators have powerful brands and extensive experience selling travel, and will have a notable effect on the landscape in the short run.

Supplier-direct initiatives, a plethora of point-of-purchase choices for consumers, eye-opening deals and the prevalence of search in every aspect of online behavior have all played a role in the erosion of online travel agency share. Yet the tour operator challenge to online travel agencies has effectively expanded the market, bringing offline tour customers online. Tour operators are pursuing multichannel distribution strategies that necessarily involve acting as online agencies selling unbundled package content as components and/or in dynamic packages. Third party content is already a part of the mix, as the vertically integrated tour operators seek to become more flexible and comprehensive in their offerings.

Online agencies have certainly gained some customers from the pool of traditional tour buyers looking for flexibility and price breaks, but that has not been their principle source of growth. So even as intensified competition has slowed growth among online agencies, the selective adoption of an agency model by tour operators (and/or their subsidiaries) will bring more product into the market and lure more hard core traditional tour buyers online. As a result, there is a relative stasis in share between online travel agencies (including tour operator businesses acting as agencies) and supplier-branded Web sites, despite the substantial advances of the latter.

Growth rates in the overall online travel market in Europe will cool to a pace that more closely resembles the revenue gains of the most forward-looking travel companies. But pan-European trends have not dominated the individual travel markets of Europe, and local market profiles vary accordingly. For example, there are no meaningful France-based LCC players, Spanish and Italian tour operators prefer to remain essentially wholesalers to online distributors, homegrown online travel agencies play significant if not dominant roles in the German and Spanish markets, and new LCC players continue to emerge even in the most mature markets such as the U.K.

Clearly, unique sets of habits, legacy players and other competitive factors exist in most country markets. Without considering these particulars, it is not possible to coherently identify trends and opportunities in the European travel marketplace. Pivotal suppliers, which increasingly hold the key to local online growth, vary from market to market. The roles and types of media involved in the travel markets often depend on the behaviors of the consumer and the historical structure of the country market.

Although the U.K., the leading and largest European market, has seen its online market share decline from nearly half of the European online travel market in 2002, it still held more than one-third share in 2006 (see Fig. 4). By 2006, Germany's share of the European travel market had nearly doubled to 21% from 2002, when it was under 12%. However, after peaking in 2005, Germany has ceded some share to the newer online markets in Spain and Italy, largely due to torrid growth rates in southern Europe.

It is intriguing to note that the U.K. and Germany, the two markets with the most profound competitive pressures, are now the two with the lowest rates of online growth in Europe. The fact that they are Europe's number one and number two travel markets, respectively, contributes to the intensity of the competition largely because of the shrinking pool of offline business to pick from: much of the low hanging fruit has now been plucked from these two markets.

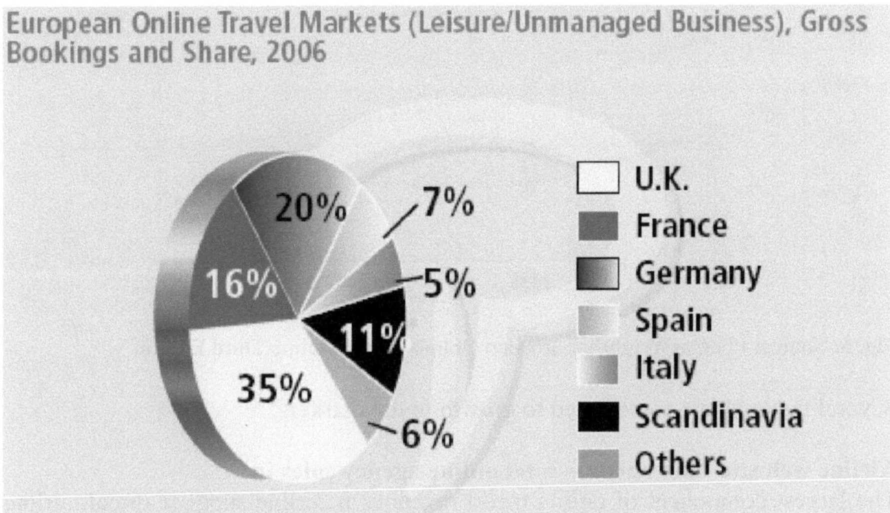

Fig. 4: Source: PhoCusWright's European Online Travel Online Third Edition

3 Online Travel Industry Growth Factors

The European online travel industry is quickly evolving. While growth varies in the different markets of Europe, there is a steady convergence of growth rates to the extent that by 2008, the five top markets in Europe will be growing within a range of 20 percentage points of the total Europe average (see Fig. 5).

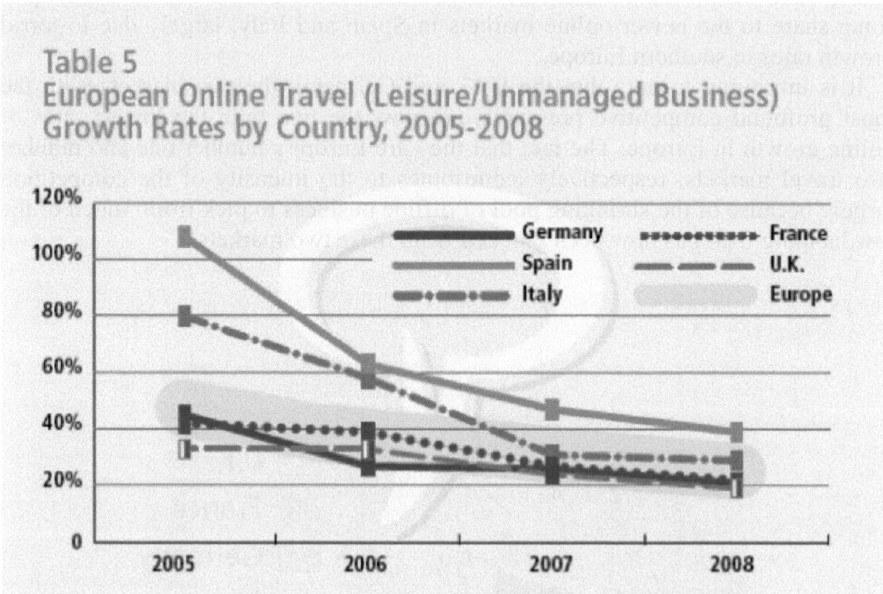

Fig. 5: Source: PhoCusWright's European Online Travel Online Third Edition

Several factors have contributed to growth in the market:

Airline web site sales surpass total online agency sales in
The largest component of online travel revenues is airline supplier direct; airline Web site gross bookings surpassed online travel agencies in 2006 with 35% share of all online travel gross bookings versus 34% for the online agencies (see Fig. 6).
Airlines lead supplier-direct charge
Airlines (LCC and traditional) dominate the supplier-direct channel with 54% of all supplier-direct gross bookings.
As markets grow, the influence of air declines
More mature markets (U.K., Germany) will witness waning influence of air in the online direct channel in coming years due to online saturation of air.
LCC effect will be repeated in emerging markets
Emerging online markets (Spain, Italy) will benefit from LCC activity bringing new Internet users to book online.
Tour operator package business makes online mark
The package (pre-packaged and dynamic) business of tour operators will grow apace with online travel agencies, and represent a substantial portion of the total online package business (see Fig. 7).
Hotels are hot and local online agencies are hotter
Hotel aggregators and local market online travel agencies will drive growth in the online travel agencies segment.

Europe has bigger potential than the U.S.

The total population of today's enlarged European Union is 457 million people in 25 countries. This is a huge market on paper, larger than North America. The original E.U., which includes the most developed countries in Europe and is therefore a more apt comparison for the North American market, involves a population that is 17% larger than North America.

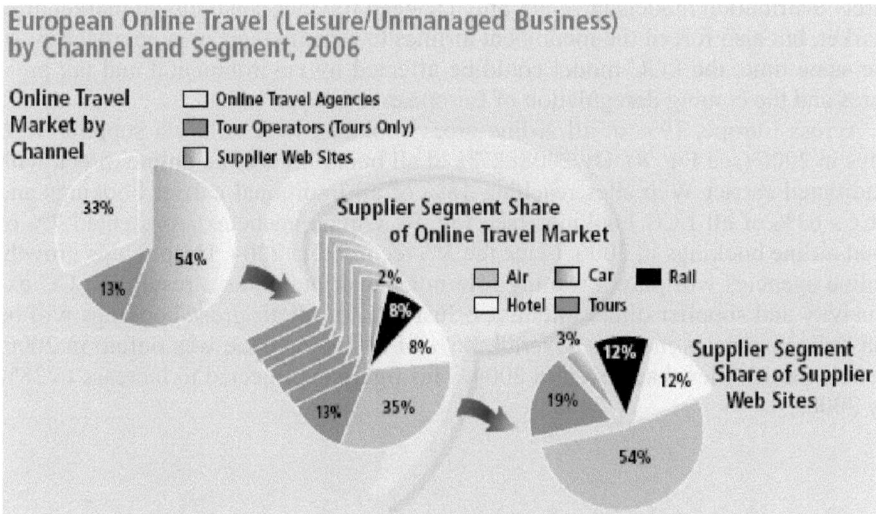

Fig. 6: Source: PhoCusWright's European Online Travel Online Third Edition

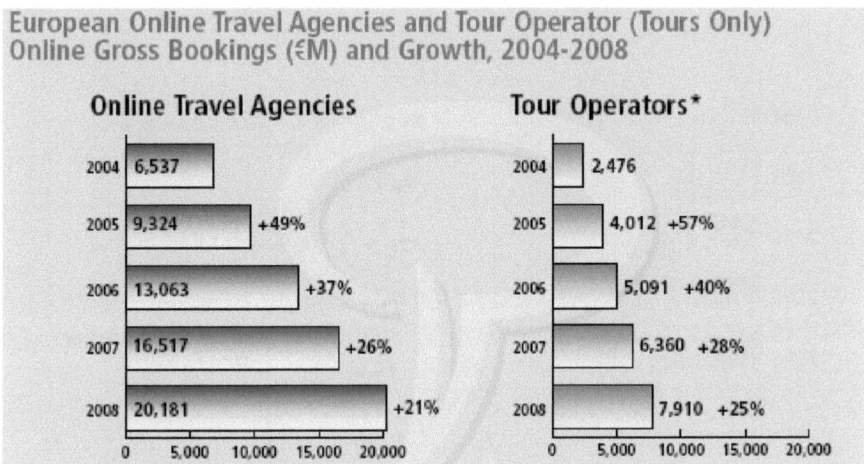

Fig. 7: * Pre-packaged tours (excludes LCCs and online travel agencies owned by tour operators); Source: PhoCusWright's European Online Travel Online Third Edition

4 Online Travel Industry Growth Factors: Airlines

The airline markets in Europe, radically transformed by the LCC model that swept the U.K. and then Germany, remain roiled by consolidation, fuel costs, huge capacity increases and the pressure to reduce capital and operating expenditures. The simplified online booking procedures and attention-grabbing deals of the LCCs' direct distribution models have not only changed traveler behaviors in market after market, but also forced the incumbent airlines to revamp their own approaches. At the same time, the LCC model could be affected by environmental and tax pressures and the coming deregulation of European railways.

Across Europe, 19% of all airline gross bookings were sold via supplier Web sites in 2006 (see Fig. 8). By 2008, 27% of all bookings will be online direct, with traditional carrier Web sites reaching 18% of all traditional carrier bookings and LCCs 63% of all LCC bookings (see Fig. 9). Online agencies represented 7% of total airline bookings in 2006, twice the 3% recorded in 2004. Despite this growth, online agencies will play a waning role in air distribution as a result of LCC exclusivity and supplier direct initiatives: Just 11% of all air gross bookings will be via online agencies in 2008. Overall, 26% of airline revenue was online in 2006; double the 13% level achieved in 2004. This figure is projected to increase to 38% by 2008.

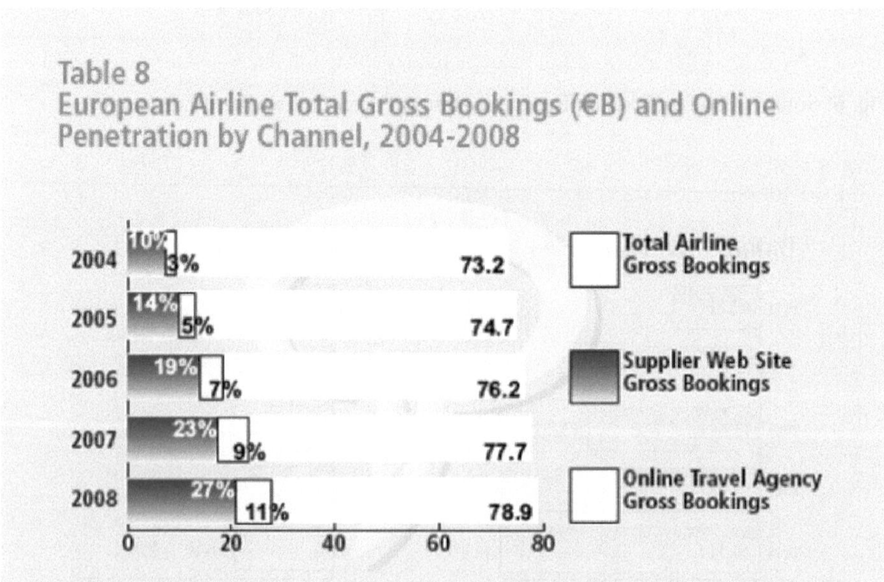

Fig. 8: Source: PhoCusWright's European Online Travel Online Third Edition

While legacy carriers have responded to market pressures by launching their own low-cost subsidiaries, the biggest effect of the LCC challenge has been a universal emphasis on supplier-direct Web initiatives. The LCC model may also encourage á la carte pricing, breaking down the basic flight product into pieces (e.g., meal, seating, baggage) more minute than the standard seat class/ticket flexibility/flight time factors that currently rule pricing.

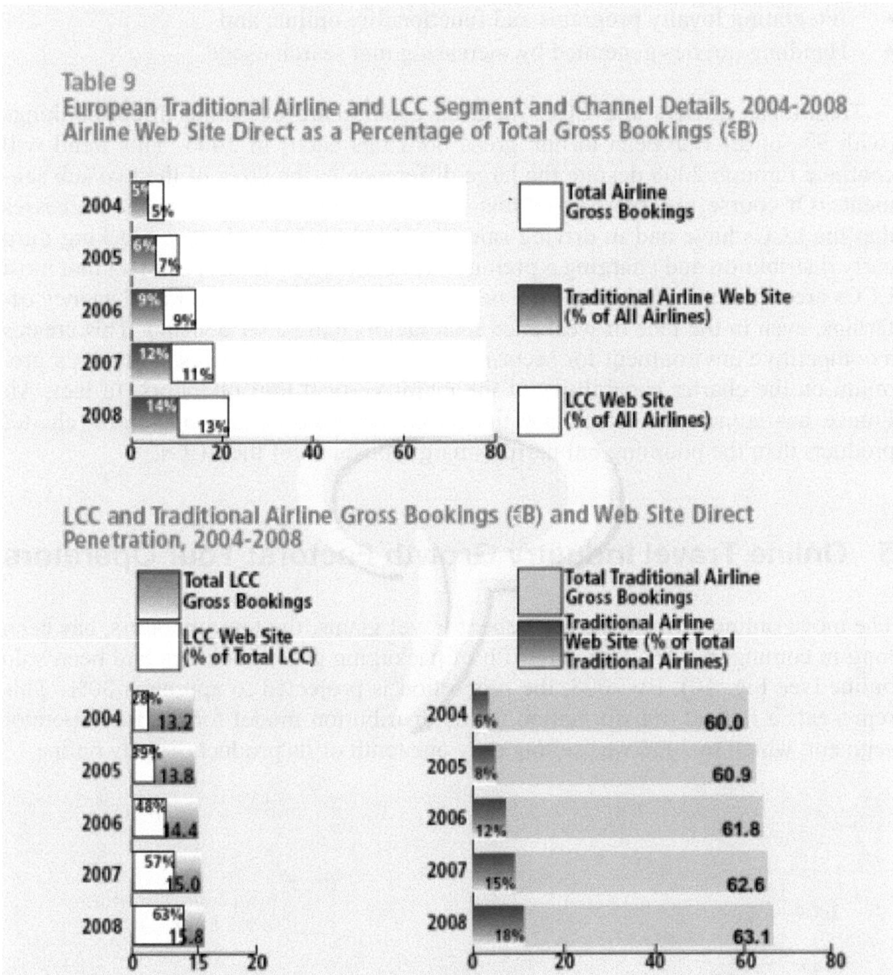

Fig. 9: Source: PhoCusWright's European Online Travel Online Third Edition

Pricing transparency and better user experience have become necessities for supplier Web sites. Airlines that have lagged in online distribution implementation and investments face numerous challenges:

- Improving usability (mainly development of a flexible, calendar-based availability and pricing engine);

- Development of business traveler sections devoted to small and medium enterprises;
- Caching data and reducing direct transaction costs in the IT and telecommunications infrastructure;
- Integrating loyalty programs and functionality online; and
- Handling queries generated by increasing met search usage.

Traditional carriers and LCCs held an equal share of online direct bookings (with 9% of all European airline gross bookings each) in 2006. This trend will continue through 2008 despite the large difference in the sizes of the two sub segments. Of course, a large part of this statistical anomaly is due to the successes that the LCCs have had in driving sales to their branded websites, blocking third party distribution and charging a premium for offline services. It is a fact that most LCCs are unwilling (and have little need) to participate in online travel agency offerings, even in the face of weakened load factors and softer demand. This creates a competitive environment for securing air for dynamic packages, and puts a premium on the charter capabilities of the major vertical tour operators. In fact, Air France has launched a subsidiary that is aimed more at tour operators' charter products than the booming but narrow-margin business of the LCCs.

5 Online Travel Industry Growth Factors: Tour Operators

The move online of Europe's incumbent travel giants, the tour operators, has been long in coming, but by 2006 one fifth of packaging gross bookings had been sold online (see Fig. 10). By 2008, the proportion is projected to approach 30%. This represents a radical transformation of the distribution model for the tour operator segment, which in 2005 was selling only one tenth of its product directly online.

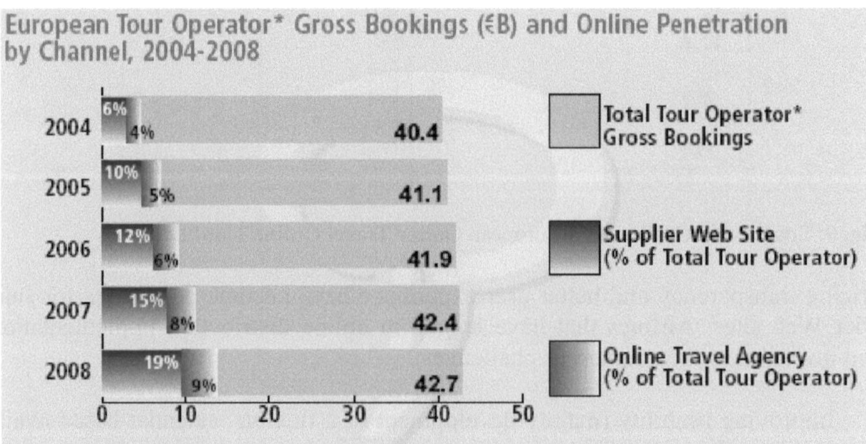

Fig. 10: Source: PhoCusWright's European Online Travel Online Third Edition

For decades a small group of giant vertically integrated tour operators ruled leisure travel across Europe, selling pre-packaged tours by high street travel agencies and through brochures. Online behaviors have relentlessly eroded that position to the point that the current behemoths, undertook radical restructuring regimens to transform their businesses with different strategies:

One strategy focused more on taking on the new challengers, online travel agencies and LCCs, with their own tactics. The second strategy was literally to shrink the businesses to focus on high-margin specialty products thought to be too difficult to distribute online.

Online challenges from LCCs and online travel agencies spurred tour operators to embrace online travel agency approaches, aiming to dynamically package their committed and/or third party inventory and to offer third party tours and travel components. But online competition is fierce and in all segments, margins are thinning.

The consolidation of the big four European tour operators is a result of sea changes in the way travelers (initially and especially in the U.K.) research, plan and purchase their travel. It is clear that LCCs have had an impact, but sometimes the scale of that impact is not always acknowledged: There has been a 20% market share shift in air passengers from tour operators to LCCs in the past eight years for travel within Europe.

In addition, chains and smaller hotels have been getting savvy about marketing direct channels, and hotels aggregators such as Booking.com have added to the pressure, empowering customers to eschew the pre-packaged product and instead pick and choose the elements of their trip. And finally, Travel 2.0 trends have helped to break the grip on information tour operator agents once had, spreading via the Internet the wealth of formerly impossible-to-find details about, say, a trip to Vietnam or a weekend in the south of France.

The advance of online information and recommendation sites and the enhanced transactional capabilities of online agencies and tour operator sites created a pervasive transparency in the planning and purchase process. This stripped substantial value from the role of the traditional travel agent, and the traditional tour operators have been forced to reinvent themselves.

6 Conclusion

Europe has not captured the same proportion of the worldwide online travel market that it holds of the total travel market. This is a basic indication of the potential that exists in the channel. However, to a large degree that will have changed by 2008, when Europe's share of all travel gross bookings in the three major regions of the world, the U.S., APAC and Europe, will match its share of online gross bookings in the regions.

Europe has bigger overall market potential than the U.S. in terms of total population. Even the original E.U., which includes the most developed countries in Europe and is therefore a more apt comparison for the North American market, involves a population that is 17% larger than North America, but the expanded EU is approaching a half billion people.

One of the most important effects of the LCC challenge has been a universal emphasis on supplier-direct Web initiatives. Further influences of the LCC model will be the encouragement of á la carte pricing, breaking down the basic flight product into pieces (e.g., meal, seating, baggage allowances) more minute than the standard seat class/ticket flexibility/flight time factors that currently rule pricing. As packaging technology becomes more sophisticated, this tendency to further unbundled components could help online travel agencies in their quest to differentiate and add value. In the U.S., the GDSs appear to be waiting for investment partners to develop such offerings. But in Europe, pressure from suppliers unilaterally implementing these structures will speed the process, with LCCs pressuring legacy carriers into following suit.

There has been a 20% market share shift in air passengers from tour operators to LCCs in the past eight years for travel within Europe. However, rapid expansion, intense fuel pressures, competitive challenges and environmental concerns will create varying drags on the rate of growth for LCCs, as will coming rail deregulation. By 2008, a showdown between the rail and LCC segments may be anticlimactic if the major LCCs succeed in diversifying their products (easyGroup spreading the easyJet model to cars, hotels and myriad other products) and modifying their model (Ryanair aiming for long haul routes across the Atlantic). Nevertheless, international rail offerings will provide a strong challenge to the short haul air industry.

The slow movement online by traditional tour operators left room for dynamic packaging growth, while the relative inertia of traditional offline distributors allowed newcomers to capture business migrating online. However, this will change significantly, especially in the more advanced online markets. The low hanging fruit is scarcer and the number of offline travelers to lure online is sparser. In fact, the logical target of the online agency or supplier is the buyer of pre-packaged tours at the high street travel agency. This will not be an easy conversion, as the traditional agency habits are strong, despite continued contractions in their ranks as tour operators increasingly focus on the Web. In addition, tour operators bring a great deal of distribution clout with them as they migrate online, with huge potential for the dynamic combination of content previously fixed on the package menu.

Trends in the Business Travel Market – Focus on Germany

Kirsi Hyvärinen, Michael Kirnberger

According to conservative estimates, half a million jobs in Germany owe their existence to business travel. This is true especially for the hotel and restaurant industry, transport companies, airports, travel agencies and other services involved in business travel. Also many jobs for example in the automotive industry, IT and telecommunications can be attributed to business travel. And millions of holidaymakers benefit from offers that are indirectly co-financed by corporate buyers.

The business travel market remains a reliable indicator of the economic situation and at the same time exerts a stabilizing effect to destinations worldwide. It does not fluctuate with the seasons and is largely immune to crises, as employee mobility is usually an important requirement for growth.

What are the key figures, developments and trends in the German business travel market? Generally speaking, as a consequence of economic recovery, the number of business trips has increased in 2006 as anticipated. But the prices for many travel services have also risen. The "VDR Business Travel Report Germany 2007 in co-operation with BearingPoint" goes into more detail. In this study – annually published since 2003 – any journey undertaken for business purposes that is recorded on the basis of expense reports counts as a business trip. The duration, purpose, distance and destination of the journey and the professional status of the travelers are insignificant here. As soon as expenditures are charged to the company as business travel expenses, they are counted in the following figures.

In 2006 one in three employees took at least one business trip. 157.8 million business trips (2005: 150.7 million) led to spending in the amount of 47.4 billion Euros (2005: 46.2 billion Euros) which translated into 148 Euros per business traveler per day (2005: 146 Euros). In comparison to a "typical" holidaymaker, a business traveler spends over twice as much.

The number of business trips in companies with ten or more employees went up 5% from 2005 to 2006. Just as the economy and employment began to pull themselves out of their low, several indicators are now showing an upward trend – with business travel activity out ahead. The number of staff members in businesses with ten or more employees in Germany has recovered somewhat (by + 1% to around 21 million), but has still not fully attained the 2004 figures.

The trend to shorter business trips is continuing. In 2006 a business trip lasted an average of only 2.2 days. The share of journeys lasting a maximum of three days was up to 84%. Almost half of all business trips do not include an overnight stay, a trend that has been on the rise since 2003. During the same period, the share of business trips lasting six or more days has been cut in half. The average length of stay has dropped from 2.6 to 2.2 days. "More business, less time", and on the average also more business trips per person, are part of the efficiency drive, which has been on the agenda of German businesses.

The total costs for business trips went up 3% to 47.4 billion Euros. Small and medium-sized enterprises (SMEs) in particular continued to pay more than their fair share in 2006, but were able to put a stop to the cost explosion: they spent 4% more money for 5% more trips. SMEs had to face slightly higher cost increases in 2006 than the market average, but on the whole their situation has improved. Those who rely on travel management are ahead of the game, because when rising travel volume is handled professionally, savings potentials can be tapped.

A year before, the double-digit cost explosion in many SMEs was caused by the full impact of the so called zero-commission policy for flight tickets. Travel Management Companies had started "to pass the buck" and SMEs′ expenses rocketed by 24%, even though the number of trips remained the same. At the same time in 2005, 38% of companies with travel managers saved at least every fifth Euro on their budget.

The average expenditures per business trip dropped in 2006 by 10 Euros.
This can be attributed to the following factors:
- another drop in the average duration of a business trip
- improved awareness of options for cost control
- new offerings and tools for cost control, especially for SMEs
- systematic utilization of the growing range of offers

Increase in Transport Spending

Transport is placing an increasing burden on the travel budget. Today, 54% of all business travel expenditure is spent on transport, a jump of no less than five percentage points in a single year. Rising costs have already provoked a reaction in road travel: three out of four cars newly licensed for company fleets are diesel vehicles.[1]

Expenditures for plane tickets and hired cars increased overall. The larger number of one-day trips contributes to a reduction in the total costs for overnight stays and thus to a proportionate increase in transport costs. The demand for plane tickets for business trips will continue to grow more strongly than the use of rail or hired cars. No wonder, as many of the growth destinations for the German economy lie far beyond the range of road and rail. This is why placing a one-sided burden on air traffic would be poison for the labor market.

---

[1] Survey by Dataforce, 2007

While an increase of 3% - 5% in flight prices and 1% - 5% for hotels was expected worldwide in 2006, it was anticipated that increases in Germany would remain below the European average. The results realized in 2006 demonstrate that cost control measures were successful in many cases.

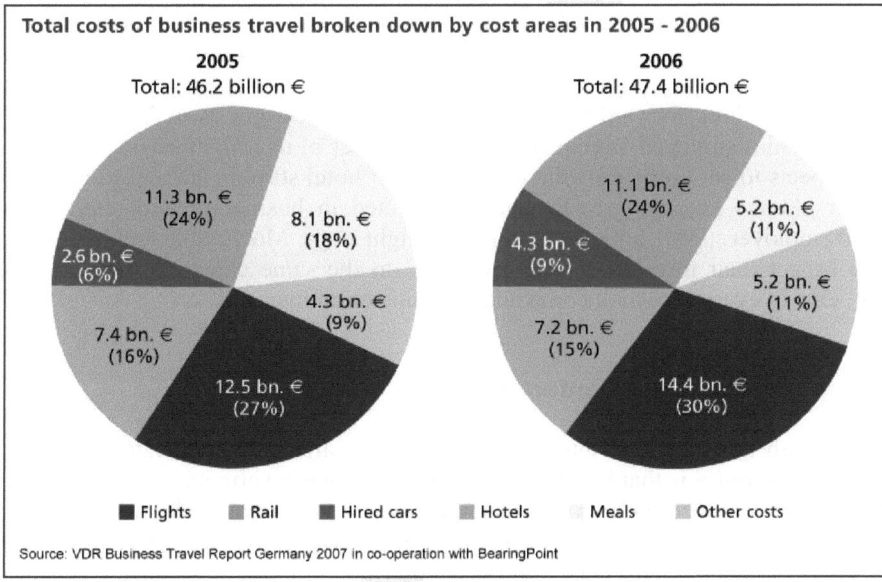

Fig. 1

Costs for Overnight Stays

In 2006, German business travelers accounted for 51.9 million overnight stays. Larger companies with more than 500 employees contributed a greater share (+3%) to the overall growth of 1%. The costs for overnight stays have remained proportionate to those in 2005 - one in every four Euros of travel spent goes toward hotels. Costs for meals however declined.

German business travelers spent a total of 40.3 million room nights in domestic accommodations in 2006, a slight increase over last year's total. Larger organizations traveled more abroad, while smaller ones increasingly stayed within Germany. The turnover of 8.6 billion Euros remained stable, corresponding to about half of the overall turnover earned by classic hotel establishments.

About 2.5 billion Euros were spent on foreign hotel stays. The average worldwide cost of an overnight stay, taking into account all types and categories of accommodations, went down 8 Euros in 2006 to 134 Euros. Possible reasons include increasing competition amongst hotels and the price transparency afforded by Internet technologies.

One in three companies is sending more employees on business trips - notably SMEs. Companies with over 1,500 employees are in particular setting off for new destinations (36%). 5% of the companies surveyed plan on fewer overnight stays. 41% of these give stricter cost control as reason. Every third company has fewer employees who travel. One in five remarked that business is not going so well (18%).

Can hotels look forward to welcoming more business guests in 2008? 70% of the companies surveyed anticipate the same number of overnight stays, but one in five expects to see some growth. The number of hotel stays is thus set to increase further. Above average growth can be expected in businesses with more than 1,500 employees: 38% anticipate more overnight stays. Most stable are the SMEs: 82% expect their travelers to stay overnight to the same extent as the previous year. One in every ten small companies anticipates an increase.

Telephone and Video Conferences

Unlike with leisure travel, business destinations cannot be substituted. Another decisive difference is that business trips can sometimes be eliminated if similar results can be achieved through alternate means. In other words: growth in the business travel sector is not a corporate goal as such. A travel manager can conduct intelligent comparative analyses, ultimately contributing to reducing CO2 emissions by exploring “virtual mobility” options such as telephone and video conferences, without the company's operating results suffering a negative impact.

In fact, to thirds of companies are already utilizing telephone and/or video conferences. A closer analysis reveals that the inclination to use these alternatives can be related to a company's travel structure. Where the ratio of foreign travel is at least 20%, the tendency to save costs - as well as CO2 emissions - by taking advantage of alternatives to business trips is greater than in companies with more domestic travel. Is the German public sector likewise setting a good example? Not quite yet: here the share is only around 40%. This is a clear signal to the political arena to take a closer look at the own potential for savings and ecological efficiency.

But traveling to meet new customers will hardly be replaced anytime soon as a guarantee for growth. This is also demonstrated by the reasons given for predicting more overnight stays in 2008. For face-to-face meetings that can be avoided, though, there are acceptable alternative forms of keeping in touch. This strategy makes sense especially for meetings that are classified as internal, regular or international.

Growing Business Travel Destinations

The annual survey conducted by VDR has confirmed over the years that the domestic travel industry profits the most from business travel. Direct industry turnover from corporate mobility in 2005 was around 36.6 billion Euros. But which cities can look forward to growth prompted by business travelers in 2008 – both in Germany and abroad?

Berlin and Munich are increasingly attractive business destinations. As a host for conferences, congresses and trade fairs, in demand not only in Germany but also abroad, Germany's capital is successfully holding its own with other international meeting destinations. Lobby work and political contacts are additional reasons why people come here.

41% of the companies surveyed indicated that they would be sending employees more frequently to at least one foreign city in 2008. London is in the lead here, but three of the first five growth cities for business travel are in the East (Shanghai, Beijing and Moscow). As a German business partner, the USA has also been a front runner especially with New York, but as well as many other cities.

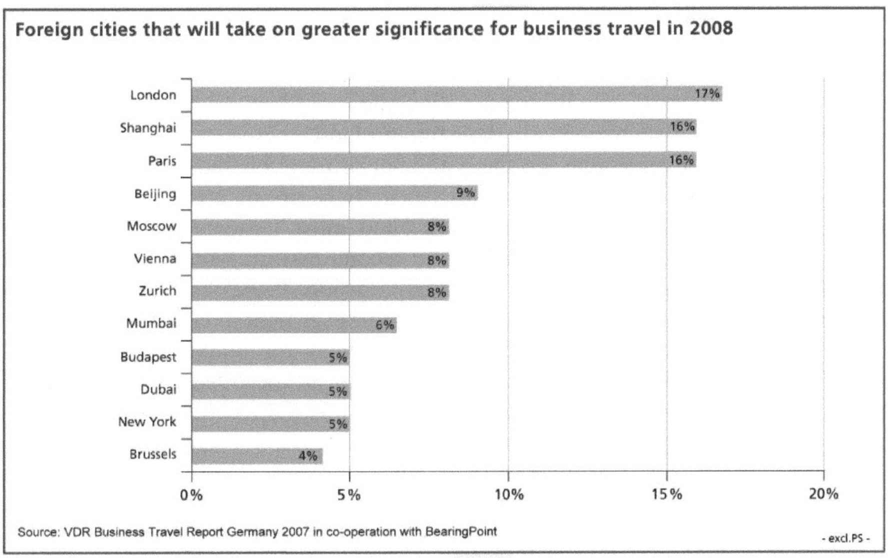

Fig. 2

Fields of Travel Management

Travel safety remains one of the top three areas that increasingly belong to the field of travel management. 37% of travel managers are involved today in "insurance/safety". In addition, 9% of the travel managers surveyed are still responsible for the three sub-areas of settlement of expenses, event management and insurance.

In 2005 the survey identified that the motive for business trips was mainly customer-oriented. In 65% of business trips, customers were met and business was negotiated. Events like fairs and seminars accounted for 35% of all trips.

While in 2006 the area of event management was increasingly being entrusted to a travel manager, a reverse trend can be observed in 2007. This shift possibly indicates that companies are aiming at a "consolidated specialization" in this field: an event manager has been placed at the side of the travel manager. Training and conference needs are experiencing a positive trend as a result of overall economic growth, with an above-average increase in meetings and congresses. As one might expect, in the SMEs only one person is responsible for both areas. 67% of travel managers here are also in charge of events, compared to 40% in companies with over 1,500 employees.

Looking Ahead

The willingness and ability of businesses to reveal their plans for the coming year has once again increased. In 2005, one in five companies was not in a position to forecast the development of their future travel picture. In 2006 this number was cut in half, and currently the basic mood is quite confident, with more companies able to make detailed statements on their future plans.

Nine in ten companies plan the same number or more business trips in the year 2008. Every fourth company will be focusing more on air travel; new fields of business and a positive order volume are cited most frequently as the reasons (78%). "More employees who travel" is the second main reason companies name for greater travel volume using all modes of transport. New business destinations will be added by every fourth company that predicts growth in travel.

The mode of transport with the lowest prospects for booking increases, i.e. where there is the least certainty of future growth, is the hired car field. Where hired cars are not needed, company cars are given as the reason, but also the airplane - even before rail travel. And only a few of those who fly less today do so for ecological reasons. The decision in favor of rail travel is likewise rarely an "environmental" one according to the respondents. Direct cost savings are still the most important factor when companies reroute or reduce travel volume.

VDR – The Business Travel Association of Germany will continue to represent the interests of German business with respect to all aspects of business travel management. The aim remains to ensure that worldwide business travel is efficient, economical, safe and obstacle-free. And the lesson for corporate leaders is clear: Companies that draw on the expertise of a travel manager are well-set to cushion cost pressure in the business travel field. Companies that ignore travel management are at the suppliers' mercy and risk their profitability.

The Future of Tourism and Travel

Future Destinations

Tony Wheeler

Thirty five years ago, when I set out from London to travel to Australia by land, I didn't realise I was about to join a travel trend. Nor did I contemplate that following that trend, and the many other travel trends that would follow, would engage me for all the subsequent years down to the present day.

1 Asia Overland

In 1972 I was 25 years old and recently married. My wife was just 22. I had just completed an MBA at the London Business School and we had decided to take a year off and 'get travel out of our systems.' In fact we achieved the complete opposite, we got travel thoroughly into our systems.

Our plan was to buy an old car in London and drive it as far east as it would go. Our elderly vehicle was so cheap we decided that if it broke down we would simply leave it by the roadside and walk away from it. In fact we managed to travel right the away across Europe to Turkey. We crossed the Bosphorus from European Istanbul to Asian Istanbul at a time when there was still no bridge across that historic channel and there was a distinct feeling of transition as we moved from Europe into Asia.

We continued east across Turkey and through Iran to Afghanistan where we sold our old car for a small profit in Kabul. From there we continued by bus and train over the Khyber Pass and down on to the plains of the Subcontinent. We crossed Pakistan, entered India and travelled up to Kathmandu in Nepal. From there we travelled back down into India and continued east to Calcutta from where we took what would be the only flight of the trip across Burma to Thailand. From Bangkok we hitch-hiked south through Thailand and Malaysia to Singapore. A ship carried us across to Jakarta in Indonesia from where we took more trains and buses the length of Java and crossed by ferry to Bali.

In Bali we hitched another ride, this time with a New Zealander on his yacht and sailed down to Exmouth on Western Australia's remote North-West Cape. More hitch-hiking took us south to Perth and then east across the Nullarbor Desert to finally arrive in Sydney the day after Christmas with 27 cents left between us.

It had been a wonderful trip, but it had also been a trip where we unexpectedly found ourselves participating in some of the most important travel trends of the last third of the 20th century.

- We were classic members of the 'post war baby boom' setting out on the sort of travel that would become a hallmark of our generation.

- We were going further that it had been typical of earlier generations. Travelling like we did across Asia would have seemed extraordinarily adventurous to an earlier generation, now it was becoming commonplace.

- We were travelling independently, another trend that would become increasingly common in the years to come.

- Many of the destinations we travelled through were still relatively untouched by mass tourism and in the following years would move from being unusual or specialist destinations into the tourist mainstream.

In fact it seemed to us that there were many other young people travelling in a very similar fashion. We did not think we were particularly pioneering and yet, it seems, we were. That trip led to the publication of the very first Lonely Planet guidebook – *Across Asia on the Cheap*. The reason for the success of that book – and indeed the reason for the success of other travel publishers who started around the same time such as:

- Phillipe Gloaguen's Guide du Routard – also with a guidebook to travelling across Asia
- Bill Dalton's Moon Publications – with their famous *Indonesia Handbook*
- Hans Hoefer's Insight Guides – with a groundbreaking guidebook to Bali

...was because we had all realised there was remarkably little travel information about these countries and that people were starting to travel widely in these regions.

2 Asia Travel

A year later, at the beginning of 1974, my wife Maureen and I started off on another trip, which would lead to another pioneering guidebook. Our first guidebook was an accident, we did not set off across Asia in 1972 intending to produce a guidebook at the end of the trip. This time we set off with a definite plan. We would produce the best guidebook anybody had ever seen on South-East Asia. In fact this would not be a great feat because nobody had ever seen a guidebook to South-East Asia. Our Across Asia trip was made in an old car, this time we travelled on a motorcycle which we used to explore the region for the next 12 months.

At the end of that year we stopped in a small Chinese hotel in Singapore for the first three months of 1975 and produced our second guidebook, *South-East Asia on a Shoestring*. That book – nicknamed the yellow bible because of the yellow cover it has always had – is now in its 13th edition and is one of many LP guidebooks which has sold over one million copies. It's important to remember how much the region has changed in the 32 years since we produced that first edition.

- Thailand was still principally known as having been a place for R&R for American Vietnam servicemen. When we travelled through the country in 1972 and then 1974 there certainly seemed, to us, to be plenty of tourists and yet the numbers were probably in the 10s or 100s of thousands. Certainly not the millions that visit Thailand today. At that time none of the hotels on the beaches at Phuket had been built.

- Malaysia was still dealing with periodic conflict with the remnants of the Malayan Communist Party.

- Singapore still had many kilometres of old shopfront places and the transvestites still appeared nightly at Bugis St. I sometimes joke that in the '70s the city state had not yet become a single vast air-conditioned shopping centre.

- Most of Indochina was either effectively off limits or about to become so, Year Zero for the Khmer Rouge in Cambodia was 1975 and the Vietnam War reached its final conclusion with the fall of Saigon that same year. It would be another 15 years before tourism to Indochina restarted.

- Indonesia, in the 1970s, was reopening for business under Suharto after being effectively closed off for much of the Sukarno era. The pioneering surfers were busy rediscovering Bali. In contrast the 1960s were the era of Sukarno's 'konfrontasi', when burning the British Embassy in Jakarta was more his style than welcoming tourists.

Those first two guidebooks, the guide to travelling across Asia and the guide to South-East Asia, were signs of places opening up. Of course many of the places we saw opening up as we travelled across Asia – like Iran and Afghanistan – have effectively closed up again. But others – like the countries of South-East Asia – have become some of the most popular tourist destinations in the world. In the subsequent years my company has grown in parallel with the growth of world tourism in that period. We have gone from a company with two people and two books to a company with offices in Australia, the UK and the USA, more than 500 staff and more than 500 different books in print. We also have a television production company, a fully digitised photo library and an extremely popular travel website.

We have seen huge changes in that time. People often point at the Internet or the jumbo jet as the biggest change in travel in that period, but I think it's equally relevant to look at the change brought about by just one country.

- When we first started producing guidebooks China was completely closed off to the outside world. We talked about going to Hong Kong and looking across the border with binoculars at the rice farmers. That was as close as most people got to China.
- Today China is rapidly heading towards becoming the world's biggest tourist destination both inbound and outbound. Shanghai is one of the world's mega cities.
- Not only do we produce a best selling guidebook to China in a variety of languages for visitors to that challenging country along with regional and city guides to China, we are also producing guidebooks to the outside world in Chinese.

3 Traveller Categories

There are numerous interesting groups of travellers, but let's start with my own ageing group, the baby boomers:

3.1 Baby Boomers

The children of the post-World War II 'baby boom' are now moving towards retirement age. It's a group with many interesting characteristics. There are lots of them and they've got more money than their retirement age predecessors. They've also got more interest in travel, they've grown up on a diet of regular travel, far more than any preceding generation. They're also healthier, more adventurous and more willing to attempt activities, which were once viewed as unsuitable for their age group. Ski resorts used to offer discounted ski passes to over-60 year olds, for example. It was regarded, at that time, as unusual for somebody over 60 years of age to still be on the slopes.

So I expect to see more travel opportunities developing for that group, more trips which require the money we didn't have when we were young, the time we didn't have when we were tied up with families and careers, but travel, which also requires the energy and spirit I hope we've retained. We've seen a big growth in activities like Antarctic tourism which fits that time-money-energy requirement. I think we'll see a lot more.

3.2 Young Backpackers

At the opposite end of the travel age spectrum are the young backpackers. Compared to earlier generations at their age, young backpackers are much more likely to contemplate overseas travel, as opposed to purely domestic travel, and when they arrive at their destinations they will find far more facilities and opportunities targeted directly at their age group and their travel inclinations.

This category of traveller has become hugely important to many countries. Their travel patterns include longer stays than other categories of travellers, so even though there daily expenditure may be much lower, their total expenditure is often very comparable to other higher spending groups.

They are also highly targeted towards activities, so while they may economise on food, transport and accommodation, they are quite likely to spend more money on activities such as scuba diving trips.

Their expenditure is often very 'sticky,' i.e. it stays in the country rather than immediately being transferred to some overseas operation. This is because they are more likely to deal with small local accommodation facilities and tour operators rather than with a major international operator. They are often travel pioneers, discovering and developing new destinations before the infrastructure necessary for more traditional traveller categories has developed.

For a travel business they are an especially important category because they will be around for many years to come. If you can gain their loyalty at the age of 20 you can look forward to their business for a long time.

3.3 The Under-30s

They're more affluent and more travel savvy than a generation ago. Compared to previous generations, they marry later, are less likely to have children and are more likely to expect travel to be part of their lifestyle. For every category of travel, from adventure travel to luxury travel, they are a very important group.

3.4 Travel with Children

The under-30s may be less inclined to have children than an earlier generation, but couples still have children and for many of them travelling with their children is an important part of their lives. We will continue to see growth in child-friendly destinations and many specific styles of travel will emphasise how they can appeal to children as well as adults. My own children are well past the 'travel with parents' stage now, but some of the best trips with my children involved activities like trekking in Nepal or safari trips in Africa.

4 'Other' Nationalities

Until comparatively recently travel was often a case of 'us' visiting 'them,' i.e. the first world travelling to the third. Rapid economic growth, which has seen many developing nations make the transition to developed status in a very short period of time – the 'tiger' economies of South-East and East Asia for example – has changed that equation.

In addition large nations, which are still part of the developing world – China and India as the prime examples – have also developed large middle classes. As a result travel has become a much more inclusive activity and in the coming decades there will be huge growth in tourism from developing nations both within their own regions and to the developed nations of Europe and North America.

Even within the developed world travel was much more popular amongst certain nationalities, in particular the English speaking nations (the UK, Australia, Canada and, to a lesser extent, the USA), plus the German speaking nations and the Nordic nations. The Japanese were big travellers, but in a very regimented, packaged fashion. South Koreans were extremely restricted until 1988, when their doors were opened at the time of the South Korean Olympics.

Amongst Western Europeans you did not expect to see the Spanish or Italians travelling in big numbers or to unusual destinations. Today you certainly do. In addition citizens of the old Eastern European nations and of the USSR were almost totally restricted to their own region. We've had a huge expansion in travel amongst citizens of that region, led, of course, by the Poles. Equally I've been surprised how many Russian trekkers I met on the way to the Everest Base Camp and Russians are now a very large part of the market for luxury travel in many locations.

Nevertheless it will be the growth of travel from China and India which will have the biggest impact:

4.1 China

China is already on its way to becoming the world's biggest tourist destination, within 10 years it will overtake France for first place. The Chinese will also become the world's most numerous tourists, but they won't fit into a single easily defined category. Already the Chinese have spawned their own local backpackers who are known colloquially as 'slow donkeys.' Although so much attention is focused on overseas travel by the burgeoning Chinese middle class, there has already been huge growth in domestic travel. The rapid expansion of car ownership in China has also led to the growth of the Chinese equivalent of bed & breakfasts, affluent Beijing families are quite likely to drive out to a village for a weekend to sample the forgotten rituals of rural life.

4.2 India

India is not far behind. We've heard regularly about the growth of the Indian middle class and we've seen Bollywood movies become mainstream. Indian tourism has always been very strong domestically, but now they're moving internationally. Interestingly, satisfying Indian culinary demands can be very demanding. Indian tour operators report that visiting countries like Italy, where the food is extremely bland to Indian tastes, can be very challenging.

5 Activity Travel

Increasingly travel is related to an activity, we're not just going to France because we love France, we're going there because we want to do a cycling trip. Sticking to the French-speaking world, we don't just go to Tahiti and French Polynesia for sand, sea and palm trees, we want to go scuba diving and meet the reef sharks. Bird watching is the focus for many tours in many countries. Some of the activity focused travel possibilities include:

5.1 Learning Travel

Education has become a lifelong activity. It's not over when we finish school or university, today we constantly have to update our computer skills or come to grips with some other new technology. And we like learning, so there's a constant interest in combining travel and learning. That can include, amongst many other possibilities:

- Cooking – French and Italian cuisine have always attracted followers, but now we can travel to Bali for an Indonesian cooking course and in South-East Asia we'll also find Thai, Cambodian and Vietnamese cooking schools. Name a great cuisine and there'll be a cooking course available. Even countries where the cuisine is not so notable – Ireland for example – seem to manage it.

- Languages – France again is an extremely popular centre for language courses but many other places also feature language learning possibilities. The Guatemalan town of Antigua, for example, is virtually dedicated to Latin American Spanish courses.

- Photography – Trips built around improving photographic skills and led by a professional photographer are very popular.

- Driving – It's possible to spend a couple of days learning to drive a Ferrari faster at the same track, where Michael Schumacher tested his Formula 1 cars. Driving courses also include 4WD courses in desert or off-road conditions or winter driving courses on ice.

Of course the learning possibilities can also create travel demand in a more traditional fashion. Many centres of further education are now heavily dependant upon oversea students, often paying far higher tuition fees than their local counterparts. There are many universities in the west, which would be facing major financial difficulties without their Chinese and Asian student numbers.

5.2 Gap Years & Career Breaks

Australians and New Zealanders have always been great believers in the 'year off,' the break between education and career. In part this is because the distance from their countries to the rest of the Western world is so great that having made the trip to Europe or North America, you might as well make the stay a long one.

- Gap Years – The British have coined the phrase 'Gap Year' and taking a break between school and university has become not only the norm for many young British travellers it has also become a major industry. Such a large percentage of the British Gap Year travellers head to Australia and New Zealand that they're targeted as a distinct category. The British may have defined the term, but there are many other nationalities where the Gap Year is becoming increasingly common, Scandinavians in particular, but the idea has yet to spread to the USA. In Britain the Gap Year has become such a phenomenon that your Gap Year travels are even something you're expected to feature in your CV.

- Career Breaks – There are gap travel possibilities at other stages as well. One of the unexpected benefits of the dot.com era in the '90s was that many Americans became much more internationally conscious and much more interested in travel because of the world-wide scope of dot.com companies. For once having an interesting gap in your CV was not something to avoid in the US. Dot.com companies were often actively interested in people who had been to strange places. With the 'job for life' becoming less common more people are taking a break between jobs.

5.3 Adventure Travel

Adventure travel has also been a strong growth area. We'll see more trips built around a trek – or a whitewater rafting expedition – or a wildlife safari. All the publicity about 'tourist ascents' of Everest shouldn't distract from the reality that these days there are a lot more tourist mountaineering ascents of every level of mountain. And plenty of other 'adventure travel' activities have become tourism mainstream as well.

- Mountaineering – Ascending Himalayan peaks was once purely the domain of large expeditions. Now Himalayan mountain climbing is open to anyone who has the ability and the money. Although media attention is often focussed on high cost peaks like Mt Everest a great deal of much lower key and lower cost, though often equally challenging, climbing also goes on not only in the Himalaya, but in the Alps of Europe, the Rocky Mountains of the USA and Canada or the mountains of South America.

- Scuba Diving – Diving has become far more popular in recent decades and is now an important activity for tourists in many more countries. When I look at my own scuba diving logbook I have been diving in 28 different countries. In some countries, in the Pacific Ocean for example, diving is a major attraction for a very large proportion of visitors. Even in countries with many other tourist attractions – Egypt or Australia as two obvious examples – scuba diving is still a very important component of the total tourist business. Many countries also have important secondary diving tourist businesses such as scuba diving schools.

- Skiing & Other Winter Sports – Despite global warming concerns, downhill skiing and other winter sports such as cross country skiing and snowboarding, continue to grow in popularity and new previously untouched areas are opening.

- Other important adventure travel activities include kayaking and canoeing, windsurfing, whitewater rafting and rock climbing.

5.4 Walking & Cycling

Once you've got to your destination, the means of travel is often as important as the trip itself.

- Walking – I'm a firm believer in the virtues and pleasures of walking. I try to spend at least a week or two on long walks every year and I think walking tourism has great opportunities for many countries. Trekking, the term usually applied to walking in the Himalaya, is one of the major tourist attractions for Nepal with walks like the Annapurna Circuit or the Everest Base Camp Trek. There are many other places in the Himalaya, which could attract many more walkers than they do. Walking trips also attract many visitors to a number of European countries, to Australia, New Zealand, the countries of Latin America, to Canada and the USA.

- Cycling – The growth of interest in competitive cycling, like the Tour de France, is mirrored by the growth of cycling as a sport and general activity. Many of these 'weekend cyclists' go on to cycling holidays, particularly in Europe where distances can be manageable, a network of minor roads allow cyclists to escape the competition of motorised traffic and there is a plentiful supply of accommodation and dining possibilities.

5.5 Wildlife Travel

Watching wildlife is, of course, one of the major reasons for visiting the 'safari countries' in Africa. In some countries the value of tourism is a key element in the continued survival of threatened wildlife species. Watching wildlife is also an important part of tourist programmes in countries from the Arctic north to the Southern lands of the Antarctic.

In some countries, such as Australia, the wide variety of wildlife is an attraction but there are also tourism opportunities for very specific species. The orangutans of Sumatra, the turtles of Malaysia, the pandas of China are all examples of this tourism field. I have even spent an evening watching badgers in England.

Scuba diving is largely tied up with watching aquatic life from small reef fish to turtles, manta rays or sharks. Opportunities also exist for swimming with whale sharks (Exmouth in Western Australia), humpback whales (Rurutu in French Polynesia), seals (Port Phillip Bay, Melbourne, Australia), dolphins (Kaikoura in New Zealand) or even, in the security of a stout cage, great white sharks (Gambaii in South Africa). Of course less active whale watching is a popular tourist activity at many destinations around the world.

Bird watching is another growth area with a great many destinations – Peru, Ethiopia, Kenya, Costa Rica, Australia, the Falkland Islands – famed for their bird watching opportunities. Specialised bird watching trips are increasingly common and a word has even been coined for obsessive birdwatchers, who are known as 'twitchers.'

5.6 Travel Routes & Marathon Trips

Following multi-country travel routes on trips, which can last for weeks or even months, include both independent and organised trips. The Silk Route from Europe through Central Asia to China is an example of a travel route, which is becoming increasingly popular and encouraging cooperation between the countries along the route. Ozbus, a pioneering attempt to follow the old Asian 'Hippy Trail' of the 1960s and '70s, departed London in mid-2007 with plans to reach Australia in three months. The South American 'Gringo Trail' and the Cairo to Capetown route are other popular routes.

Organised travel along these long routes include:

- Tour d'Afrique – an annual three month bicycle ride from Cairo to Capetown.

- Plymouth-Banjul Challenge – an annual three week drive, starting in England and concluding in Banjul, Gambia. Entrants purchase older vehicles for the event and at the conclusion the finishing vehicles are auctioned off with the money going to Gambian charities.

- Train trips – probably the best example of the long train trip is the Trans-Siberian Express (or the Trans-Mongolia and Trans-Manchurian offshoots), but there are many other short and long train trips ranging from the north-south Ghan in Australia to the Blue Train in South Africa.

- Motorcycling trips – organised motorcycle trips include a variety of Australian outback tours, trips in other remote areas or trips around India using brand new, but vintage style, Enfield India motorcycles.

5.7 City Breaks

To some extent 'city breaks,' making very short trips centred around a visit to some city, are still a predominantly European phenomenon, driven by the proliferation of low cost carriers (LCC), which have revolutionised air travel in the past decade. The European no-frills airlines derive much of their income from the 'weekend-away' business. In Europe, the popularity of Eurostar has prompted many Londoners and Parisians to cross the channel for a weekend away. Reykjavik in Iceland has become a European party town on summer weekends when the midnight sun is on call.

One of the determinants of growth in this field, however, has to be smooth transitions and lack of red tape. People are not going to travel somewhere for a short trip if it involves getting a visa or spending hours getting through immigration, retrieving baggage or travelling into town. Nevertheless with the growth of low cost carriers in Australia and Asia, city breaks are likely to grow in other areas outside Europe.

Totally unplanned short trips are, at present, also a predominantly European phenomenon – the LastMinute.com experience. This can even involve going to the airport, seeing what is on offer from charter companies and then selecting the best deal for an immediate departure.

5.8 Cruises

Ship cruises are enormously popular and growing in popularity. Although at one extreme cruising seems to be the opposite of travel, simply staying on board a ship, which is effectively a floating resort, there are also many other cruising categories. This can vary from boutique size traditional cruise boats to a wide variety of smaller ships for more adventurous and more unusual cruises such as to remote islands of the Pacific or to the Arctic and Antarctic. Charter operations are also experiencing strong growth.

River travel is another important cruise category, whether it is the Rhine or Danube in Europe, the Irrawaddy or Mekong in Asia, the Murray in Australia or the Amazon in South America.

5.9 Luxury Travel

An expensive hotel is no longer necessarily also a big hotel, there has been a great deal of growth in small boutique resorts and hotels or in the resort segment typified by groups like Aman Resorts or Banyan Tree Resorts. There is an important and growing category of people who expect five-star comfort but don't want it in a faceless, big hotel package.

There is a great deal more specialist luxury travel ranging from chartered boats and ships to aircraft. To some extent this is driven by the expanding number of experienced travellers, who have effectively 'been everywhere.' To cater for their curiosity, travel is offered to more and more remote locations, places which are often difficult to reach or require specialist transport.

6 Travel Niches

Styles of travel are increasingly dominated by niches. The old fashioned package tour will continue to decline in importance, in part because today's increasingly sophisticated travellers are more confident about making their own bookings, finding their own way, doing things without the protective shelter of a group. Package tours will, however, be very important for the first generation of tourists from new providers like China and India.

Apart from major travel trends, a great number of interesting small travel niches have developed in recent years. For example there is:

6.1 'Volontourism'

This coined word describes travel which combines elements of voluntary work with tourism. Travel often inspires philanthropic interest and the relatively recent growth of 'voluntourism' is an indication that people want to give their time and effort as well as their money. There is a concern that voluntourism can do more for the voluntourist's conscience than it does for the place concerned, but there is no question that tourism and philanthropy can be linked. The huge international response to the countries hit by the Asian tsunami was in part because so many people had visited the affected countries and did not see them purely as anonymous nations in a news report.

6.2 Dark Tourism

'Dark Tourism' can be defined as visiting dangerous places, but it equally accurately applies to tourism to war sites or to sites of disasters, catastrophic or tragic events. I have made my own dark tourism tour, visiting the countries the US President George W. Bush labelled as an 'evil axis' which, along with six other 'misguided' nations, I wrote about in my book *Bad Lands*. Some important dark tourism sites include:

- The Cimetière du Père Lachaise is one of the most popular tourist attractions in Paris. Chopin, Balzac, Wilde, Proust, Modigliani and Piaf are all buried there, but it's rock icon Jim Morrison, who attracts the big crowds.

- Ground Zero, the site of the World Trade Center buildings in New York City, has been one of the most important dark tourism locations ever since the 9/11 attacks.

- The Catholic and Protestant murals, which decorate buildings and walls in Belfast recall the events of the long struggle between extremist groups in the Northern Irish capital.

- The Anne Frank House in Amsterdam and the concentration camps in a number of European countries recall the events of the World War II holocaust.

- Grūtas Park, also known as Stalin's World, at Druskininkai in Lithuania, features an excellent collection of Soviet era statuary. I was deeply disappointed that the post invasion chaos led to the destruction of so many fine examples of Saddam Hussein statues, they would have made a fine Baghdad tourist site.

- Other examples include the 'Topography of Terror' at the site of the Gestapo Headquarters in Berlin, Robben Island off Cape Town, where Nelson Mandela was imprisoned for a quarter century, or even Gracelands, Elvis Presley's former home, where bad taste intersects with dark tourism.

6.3 List Ticking

There can be no finer example of the enthusiasm for making up a travel list and ticking it off than the Travelers Century Club in America, which produces its own idiosyncratic list of all the countries in the world and requires potential members to tick off 100 of them. The huge success of the book *1000 Places to See Before you Die* and the numerous spin off titles that followed the original is a clear indicator of the enthusiasm for this genre of travel.

6.4 Slow Travel

Concerns about sustainability and the effects of travel on climate change has focused attention on 'slow travel' whether it is travel by train rather than air or simply concentrating travel on a small region rather than attempting to cover larger areas at higher speed.

6.5 Gambling Tourism

Las Vegas is the prime example of gambling tourism, where a host of tourist activities are built up around the central interest of gambling. To some extent the growth of gambling tourism has been based on the provision of the activity in one area when it is banned or circumscribed somewhere else. The explosive growth in gambling in Macau, for example, is based on the pent up demand for gambling outlets from Mainland Chinese. Some observers feel that as gambling opportunities become more and more widespread the demand will eventually tail off. That level has not yet been reached, however, and Singapore is the latest country announcing plans for the construction of huge casino and gambling facilities.

7 Travel Information & Influences

The provision of travel information can serve both as an inspiration and a facilitator for travel experiences. This can include:

7.1 Television, Magazines & Newspapers

Print and broadcast media are a vitally important factor in the provision of travel information and inspiration. Despite overall declines in the print media business, travel magazines are still proliferating and often appearing in new and specialised forms.

7.2 The Internet

The growth of the Internet as a travel medium has had a huge impact in all areas of travel. Internet bookings are a major factor in the decline of travel agencies and the ability to handle airline bookings via the Internet has also been a major factor in the growth of low cost carrier (LCC) airlines. For travel publishers, the provision of information on the Internet is both a threat and an opportunity. Of course guidebooks are still a very convenient way of carrying information, but increasingly the Internet is turned to as a primary source of information.

7.3 Word of Mouth & Personal Experience

World of mouth and personal recommendations still play a vital part in the provision of travel information, particularly in the growth of new areas of travel or new destinations. Today, this is also supplemented by 'user created' information on websites.

Information can also inspire people not to travel, to stay at home. Concerns about safety and travel difficulties, due to terrorism threats or the bureaucracy and inconvenience of increased security, can easily deter people from travelling. Government travel advisories are frequently cited as being a seriously negative factor for travel plans.

8 The Travel End Game - Sustainability

The ideals of sustainable travel and ecotourism have become fashionable, but serious questions are being asked about travel's impact on the environment and its sustainability. Any form of travel has an environmental cost, but air travel is particularly damaging and although its overall impact is, at present, quite limited, the carbon emissions contribution of air travel is growing faster than any other category.

8.1 Travel Guilt & Climate Change

Particularly from the UK, where the low cost carrier and city breaks phenomenon has been particularly strong, the concept of 'travel guilt' has become widespread. Just as driving a car with poor fuel consumption and high carbon emissions is coming to be seen as anti-social so is air travel, particularly travel that is viewed as frivolous or excessive.

Travel guilt could cause people to rethink their travel plans and reduce their travel consumption, but even more likely is government policy to restrict or discourage travel by taxes and other applied costs such as requiring financial contributions to carbon matching schemes.

8.2 Peak Oil

Self-enforced restrictions on travel, through travel becoming seen as anti-social, and government limitations, through taxation or other charges and restrictions, may limit the growth of travel or even reduce the amount of travel, but some futurists predict the arrival of ‘peak oil’ may have much greater and more conclusive effects.

The peak oil concept predicts that at some point in the near future oil production will reach a peak and then rapidly decline. At the moment, while oil reserves are declining we are also rapidly increasing our consumption, both through increasing affluence in the West and huge growth in the developing world. Peak oil proponents suggest that airlines and aircraft manufacturers are predicting massive growth in air travel without considering the possibility that air travel may simply become extremely expensive, due to rapidly diminishing oil reserves.

Alternative forms of energy can be used for almost any other travel energy consumption. We can contemplate driving electric or hydrogen powered cars or travelling in electric trains or nuclear powered ships, but at present there is no feasible alternative to fossil fuels for air travel.

Space Travel

Dana Ranga

Interview with Gerhard Thiele, Head of the ESA Astronaut Division ESA/EAC

Gerhard Thiele is a physicist and astronaut. After completing secondary school in 1972, he served four years with the German Navy, first as a weapons officer, then as operations officer on fast patrol boats. Thereafter, he studied physics and astronomy in Munich and Heidelberg and completed a doctor's degree on global ocean circulation at the University of Heidelberg. Gerhard Thiele then continued his research at Princeton University. In 1986, the former Deutsche Forschungs- und Versuchsanstalt für Luft- und Raumfahrt (German Aerospace Center) placed an advertisement for career opportunities for scientific astronauts in all German newspapers. Five lucky finalists were chosen from among almost 1,800 applicants – one of whom was Gerhard Thiele. Following the astronaut training program in Cologne, he was assigned to the German D2 Mission in 1990. He was initially in the backup team. During the mission, he was responsible for radio communication with the astronauts as part of the ground control team.

Twelve years after joining the astronaut team, in 2000 Gerhard Thiele journeyed with the shuttle into space. His mission was to map the Earth's surface with a radar system. The result was an almost complete three-dimensional digital model of the Earth with accuracy unseen until then. Following this mission, he became the first European astronaut to be appointed to the CapCom at NASA. In this position, he was responsible for communication between the shuttle crew in space and the mission control center in Houston. In 2001, Gerhard Thiele returned to the Astronaut Center in Cologne, where, several years later, he became the Head of the ESA Astronaut Division.

In the conversation held during the ITB 2007, Gerhard Thiele spoke about the relevance of space travel, the reasons for these journeys, and what the future has in store for this sector. Traveling has had a very large impact on the cultural development of our civilization – soon a growing number of trips will be destined for space. One of the most important steps for this development will be to prepare people for such trips.

Dana Ranga: What impelled you to journey into space, what inspired you?

Gerhard Thiele: My desire to journey in space began when I was still a young boy. I was born in the fifties, it was a different era. My parents bought their first television in 1965, when I was almost 12 years old. We had had the television for just one day – or maybe it was one or two weeks – when they began broadcasting the take-off of Gemini 3 over the television. In today's linguistic slang, it was simply *cool* to see these men stride to the rocket in their white suits with their cooling set dangling from their hand. It looked a little awkward, but it was nevertheless fascinating; I was simply amazed and I probably responded just like very many other young boys and girls. And, it was at this point where my desire to become an astronaut emerged. Adults reacted just as you would expect: They said "Keep on dreaming…" But as you see, in the end dreams can and indeed do come true.

Dana Ranga: You were in space and saw the Earth from above. The term "overview effect" is a unique experience for space travelers, which occurs when they view the Earth from such an extraordinary angle. This experience produces a certain change in the awareness of the beholder; each space traveler however experiences it on a uniquely personal level. How would you describe your first view of Earth? How did it affect your level of consciousness?

Gerhard Thiele: The "overview effect" is naturally a very complex phenomenon. I do not think that viewing the Earth for the first time is something that immediately and intrinsically affects people, or at least not me personally. The first time I was able to see the Earth, eight and a half minutes after take-off, I had just one thought in my head: You have to take a photo of the tank! I refer to the large external tank, which is mounted on the shuttle for take-off. I had to keep the tank in eyesight, get the camera, check all the camera settings, and even capture it on film all at the same time. The Earth naturally fell into my visual sight and I remember exactly what I thought at this precise moment; it was nowhere near poetic, I simply remarked: "It is just like in an Imax theatre." Through the window, the Earth looked exactly like it would if I had been sitting in the movie theatre. It took awhile before I understood where I actually was and what I was actually seeing.

To describe the "overview effect", I would like to use an analogy, which everyone has experienced. I like to go to museums and galleries and some of my favorite artists are impressionists, Claude Monet or Alfred Sisley in particular. Impressionism is a genre, which has always inspired me. If you stand really close to a painting, you can actually see the individual brush strokes made by the artist, and you have the impression that every single stroke is exactly where it is supposed to be. But if you just look at the brushstrokes, you are unable to contemplate the painting as a whole. You can only really see the whole painting when you step back to sense its completeness. The same goes for an overview of the Earth. What we see on Earth is more on the lines of a local cosmos, for me it is Bonn, or the Greater Bonn area; when I think back to my roots it is Swabia. But it is a single brush stroke on the painting. The Earth as a whole can only be experienced if you step away from Earth – and that is only possible with space travel.

Dana Ranga: When people, who have not been in space, look at the Earth out the plane window, they are often filled with a type of elation. People become aware of the different dimensions within the continuity. What is the difference between this and the awareness internalized in space, aside from the distance? Does weightlessness add another aspect?

Gerhard Thiele: Orbiting in space definitely adds a few other aspects to awareness. It is difficult to rank them and identify which one is the most intense. Every single person, who has experienced this wonder, will respond uniquely. I strongly agree with what Reinhard Furrer (a former German astronaut, R.F. passed away Sep 1995) described. When you take a plane, you are naturally far above the Earth's surface, planes fly around 12–13 km above the Earth's surface; houses are very small, at best you are able to identify certain large objects and you feel as if you are moving above the Earth, but you are not actually detached from Earth. If no one does anything, the plane will eventually run out of fuel, at which point the plane has to land. You are still an intrinsic part of the Earth since you inevitably have to land at some point. When you are orbiting the Earth, it is a little different. When you are in space, you then see the Earth below and it doesn't matter if it is 200 or 500 km away, if you do nothing, you will never return. You stay exactly where you are and that is definitely a special quality. In other words, you are no longer part of the Earth! Although you are still within the Earth's gravitational field and the forces, which the Earth exerts on the shuttle in return compels you to orbit around the Earth, you are somehow detached from the Earth. For me, the Earth was like any other celestial body, which could have been anywhere. I observed the Earth from all possible angles because the shuttle changed its position in space several times during our mission. Sometimes the Earth was in the upper left part of the window, then in the left bottom area and sometimes it disappeared completely out of the visual field of the window. It did not seem strange to me at all that we could not see the Earth; it had simply stopped being a part of my fundamental existence. Reinhard Furrer described this notion perfectly as he mentioned that the voices I hear over the head set from Earth is something that I only know. They could actually be coming from anywhere.

Dana Ranga: I imagine that while you were in the shuttle, you must have closed your eyes at least once and simply floated. What were you consciously aware of in this precise situation?

Gerhard Thiele: This is indeed a very fascinating phenomenon. I had been looking forward to this experience with enthusiasm because I had heard about it from other astronauts and had even received some tips. Here on Earth, we purposefully walk erect. If we were to walk slouched over, our back would eventually begin to hurt. Our bodies instinctively recognize when we are not optimally balanced in respect to Earth's gravitational pull. There is no such equilibrium in orbit! You quickly assume a posture in which all the muscles in your body completely relax: your shoulders tend to slouch forward; your back is curved slightly; your knees slightly bent and your arms hang at shoulder height. In this position you are in a state of complete relaxation. When you go to sleep for the first time, it is hard to fall asleep because you initially want to sleep stretched out in the sleeping bag just as you would if you were lying on your bed at home – You are on the wrong track completely! How do you wake up in the morning? In the shape of an "S": feet hang weightlessly towards the floor, knees knock on the ceiling of the small bunk, your bottom hangs near the surface of the bed while your head and hands knock against the top of the bunk. And, the second time you head off to bed, you adopt this S-shaped posture and float directly into the sleeping bag – within minutes you are fast asleep. What happens when you assume this fully relaxed position and then close your eyes? I have to admit that it only happens at night, when you are on the night side of the Earth, and when there is no stimuli, particularly when there is no light. When you close your eyes and you no longer sense your body, because you are completely relaxed, then you are simply spirit. You instinctively experience the elusive "I think therefore I am" philosophy. You are unaware that you have legs and arms or that you have a back – it becomes utterly impossible to perceive your body. The only thing you perceive is that your mind is still working. It all comes down to the mind and spirit.

Dana Ranga: Was anything different than expected, in regard to observing the Earth from a distance? Were there any surprises for you while you were up in space?

Gerhard Thiele: There are many things that is completely different. The aspect, which astounded me the most, was how easy it to move and perform tasks in zero gravity. An unhindered lightness of being truly does reign up in space. We mapped the Earth with a radar system and also captured accompanying photographic images. To achieve the latter, we actually had to go to the windows and wait for the perfect moment, when the target emerges in our field of vision. The first day or two, I kept a firm hold on the window until Janet told me: "You don't have to hold on, you simply stay where you are!" And this is such a novel concept that you have to get accustomed to it. Although it is easily said, it takes time before you actually internalize this experience.

Dana Ranga: It almost seems as if your posture in zero gravity lies within oneself…

Gerhard Thiele: That is true. Each person has their own stance and there is no other reference than oneself. The shuttle does however provide certain orientation, which corresponds to the Earth's gravitational field. If you were to enter the shuttle or visit a model of the shuttle, you would notice that the labels are exactly as you would expect, that is to say right side up. However, in zero gravity, there is no such thing as "right side up". Up is where your head is and down is where your feet are! During the first few days, you tend to adopt a "natural" position based on the equipment and the orientation of the labels. When you travel in space for the first time, you often spend about two or three days in such a position. However, somewhere along the line you suddenly realize that for all intents and purposes you seem to be standing on your head or have taken on a different position in the room, possibly because adopting such a posture permits you to complete a task with much more ease. It simply occurs without much actual thought. This is the point of no return, when you actually become a part of zero gravity! Thus it is as if each person defines their own coordinate system. A short story can perhaps accentuate this concept.

There were six people on board of our flight and we worked in two shifts day and night. After finishing our working day and handing over our "red" shift to the "blue" shift, we met for dinner in the mid deck of the shuttle. (The mid deck is located below the flight deck, which is at the top of the shuttle, in reference to E-arth's concept of orientation.) Aside from the many storage rooms and electronic equipment, the mid deck also houses the kitchen and sleeping quarters. Kevin, our commander, was already in the mid deck – he was apparently standing on his head. As Janet and I arrived, we adopted the same position as Kevin: the ceiling was now our floor and the floor our ceiling. We floated in circles, hooked on to the loop on the ceiling and were eating quasi upside down. Our colleagues from the blue shift, who had taken over the radar measurements, could not locate something, most likely because we had laid it down in the wrong place. Suddenly, Dom, our pilot, stuck his head through the hatch, which connects the mid deck and the flight deck, and asked: "Guys, where is the procedure so and so?" Kevin, Janet and I all automatically looked towards where for us would have been up, where we thought the hatch would be. But we all ended up staring in the same incorrect direction. In reality, the hatch was at our feet. And we were there wondering where Dom actually was? Then his voice arose from the opposite direction – "I am here!" This relates quite accurately to how your perceptions and references take hold in this environment.

Dana Ranga: As a CapCom team member, you were responsible for communication between the space team and the control center. How can you describe things that others have not yet experienced? How do you describe something as unique as zero gravity?

Gerhard Thiele: Communication is always a process. Naturally, I try to describe things as accurately as possible about my journey into space; however, I continually have to face the fact that words cannot adequately describe my experience. I wish I were a poet and could fashion a poem of lyrical imagery to express my experiences. Luckily, experiences often recover tangibility when you tell of them, you almost relive the experience. Being the windows to the soul, your eyes reflect your internal emotions as you speak, making the account that much more authentic for the listeners. The chairman of the German Aerospace Center recently confirmed this when he told me that astronauts' eyes light up as soon as they start talking about space travel. In light of all this, I do hope that my family and friends have been able to form an idea of my space flight because they participated in it to a certain extent in that they were interested in my journey.

I can only hope that the images we bring and try to describe are able to communicate what we experienced. This objective might not actually be so difficult:

Imagine I were to meet and converse with a randomly selected person somewhere in the world, I would most likely not be able to explain the rocket equation – unless this person were to have specific knowledge in physics. But I would most definitely be able to speak about the fascinating view of the Earth or the stars! Simply because everyone understands feelings and sensations. Everyone who has observed the night sky or watched the sun set has asked the fundamental questions: Who am I really, where do I come from, where am I going and where does this cosmos in which we are nestled lead? These are fundamental questions, which move every person. Space travel can significantly contribute to finding the answers – I must specify that I refer to space travelers not to automated probes.

Dana Ranga: How did you feel when you saw the blue and white Earth on one side and the blackness of space on the other – within the extraordinary environment of zero gravity? Which do you find more intriguing?

Gerhard Thiele: I find it difficult to answer this question because my wife is sitting among the audience and might have mixed feelings about my answer. Astronauts give completely different answers to this question. Dave Brown, who lost his life in the Colombia accident, had yearned to leave Earth as an astronaut and to fly into space. However, this desire was nothing compared to his desire to reach the blue star, as if he were from another planet. This was not the case for me; perhaps I did not spend enough time in space. But, after eleven days in space I was simply not ready to come home. To the contrary, the blackness fascinated me. I have a nice anecdote concerning this attraction:

It was our last full day of work in space. The red team had just finished working and we turned the control over to Dom, Janice and Mamoru. For some reason, God must have been at work on this day, Mamoru did not wake up right away. Dom and I naturally let him sleep some more, there wasn't much work left and all preparations had been completed within the hour. The moon came up and I turned to Dom: "So, show me what you can do, take me to the moon." For a short moment, Dom seemed a little irritated, until he saw the sparkle of mischief in my eyes and told me: "I would love to, but I don't think we can do it with this shuttle." Of course, he did without a doubt have a point: It is impossible to fly to the moon with a shuttle. Instead we installed a small stereo in the flight deck and listened to Beethoven's 9th Symphony as the Earth passed by below us. It was simply amazing… And as I looked out the window, I wondered what it would be like to actually leave Earth. I will not be granted such an opportunity, but other people will venture forth in the future. It would be nice, to experience this sensation!

Dana Ranga: Mr. Thiele, thank you very much for this conversation.

MIX
Papier aus verantwortungsvollen Quellen
Paper from responsible sources
FSC® C105338

If you have any concerns about our products,
you can contact us on
ProductSafety@springernature.com

In case Publisher is established outside the EU,
the EU authorized representative is:
Springer Nature Customer Service Center GmbH
Europaplatz 3, 69115 Heidelberg, Germany

Printed by Libri Plureos GmbH
in Hamburg, Germany